ONLY ONE SQUADRON WAS PERMITTED—UNDER DIRECT PRESIDENTIAL ORDERS.

I WONDER WHAT'S UP?

GREEN MAGNET LEADER TO M.E.K.1— REQUEST PERMISSION TO BOARD.

SHORTLY, ON M...

GREETINGS, COLONEL.

GOVERNOR!

SONDAR, OLD FRIEND! WHY THE FIGHTERS?

TO REFRESH MY PILOT SKILLS, I WAS TESTING THE NEW GREEN MAGNET SQUADRON PERSONALLY. WHEN NEWS OF YOUR ARRIVAL CAME IN I DIVERTED OVER TO OFFER ASSISTANCE.

DON HARLEY

SOON A CONFERENCE IS UNDER WAY.

... SO WE NEED TO PACIFY THE RED SHIP AND EXTEND THE SALVAGE OPERATION. SUGGESTIONS, GENTLEMEN.

OUR DEEP SPACE CRAFT WERE LOST IN THE ROBOT CONFLICT, COLONEL.

THE GREEN MAGNETS HAVEN'T THAT RANGE.

YOUR SHIP HAS CANNON— WHY NOT USE THEM?

VANDALS! HEAVEN KNOWS WHAT KNOWLEDGE THAT SHIP HOLDS — IT NEEDS INVESTIGATION, NOT ANNIHILATION!

GUNS MIGHT PROVOKE ANOTHER HOSTILE RESPONSE. NO, WE MUST BOARD AND SHUT HER DOWN.

DAN! ASTRONOMY BRANCH HAVE NOW CALCULATED THE ORBIT OF THE SARGASSO SEA OF SPACE...

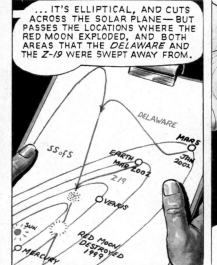

... IT'S ELLIPTICAL, AND CUTS ACROSS THE SOLAR PLANE— BUT PASSES THE LOCATIONS WHERE THE RED MOON EXPLODED, AND BOTH AREAS THAT THE DELAWARE AND THE Z-19 WERE SWEPT AWAY FROM.

DELAWARE

MARS
JAN 2002

SSofS

EARTH
MAR 2002

Z/9

VENUS

SUN

RED MOON DESTROYED 1999

MERCURY

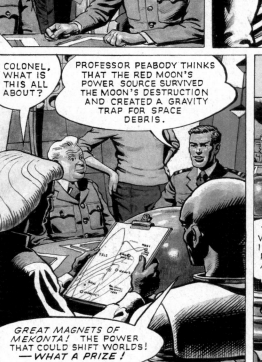

COLONEL, WHAT IS THIS ALL ABOUT?

PROFESSOR PEABODY THINKS THAT THE RED MOON'S POWER SOURCE SURVIVED THE MOON'S DESTRUCTION AND CREATED A GRAVITY TRAP FOR SPACE DEBRIS.

GREAT MAGNETS OF MEKONTA! THE POWER THAT COULD SHIFT WORLDS! — WHAT A PRIZE!

BUT, IN A SECRET LAIR ON VENUS, AN EAVESDROPPER TAKES NOTE.

YES, YOU FOOLS, WITH SUCH A POWER I COULD REGAIN MY RIGHTFUL PLACE — MASTER OF THE UNIVERSE!

CONTINUED

First published in June 2013

A catalogue record for this book is available from the British Library

ISBN 978 0 85733 286 8

Library of Congress catalog card no. 2013932251

Published by Haynes Publishing, Sparkford, Yeovil, Somerset BA22 7JJ, UK
Tel: 01963 442030 Fax: 01963 440001
Int. tel: +44 1963 442030 Int. fax: +44 1963 440001
E-mail: sales@haynes.co.uk
Website: www.haynes.co.uk

Haynes North America, Inc.,
861 Lawrence Drive, Newbury Park, California 91320, USA

Printed in the USA by Odcombe Press LP,
1299 Bridgestone Parkway, La Vergne, TN 37086

Author Rod Barzilay
Illustrator Graham Bleathman
Commissioning Editor Derek Smith
Copyediting Ian Heath and Graham Bleathman
Designer Richard Parsons

All illustrations by Graham Bleathman with additional art by Don Harley and Tim Booth.

Dan Dare comic illustrations from the *Eagle* by Frank Hampson, Don Harley, Frank Bellamy, Bruce Cornwell,
Desmond Walduck, Eric Eden, Harold Johns and Keith Watson, plus Dan Dare studio team members Greta
Tomlinson and Joan Porter.

Further artwork and illustrations from Spaceship Away by Don Harley, Keith Watson, Graham Bleathman,
David Pugh and Tim Booth.

Eagle frame scanning and tidy-up by Desmond Shaw.

Author's acknowledgements
This book is based on the creations of Frank Hampson and his team. On works published in *Eagle*, the Wallis Rigby
"Presso" *Dan Dare Space Ship Book*, *Spaceship Away* Magazine, and further developments put forward by Eagle
Society members. To fill in the gaps I have tried to blend previous thoughts from Adrian Perkins, David Ampleford,
Denis Steeper, Tim Booth, Graham Bleathman and Roy Henderson, along with my own. And I would also like to give
a huge thanks to Desmond Shaw for his superb frame scanning and touch-up work, along with Tim Booth for adding
new artwork when no frames were available. Lastly, a salute to Sue, my long-suffering wife, who became a Dan Dare
widow for most of the time I was scribbling away!

Illustrator's acknowledgements
Desmond Shaw for his invaluable help with the preparation and scanning of both the *Eagle* art and mine,
Ian Heath for additional editing, Dolby and Pixel for art supervision and Katie Bleathman for tea, sympathy
and cat wrangling!

**The adventures of the
original Dan Dare continue
in *Spaceship Away* magazine.
  For details visit
www.spaceshipaway.org.uk
or write to 8 Marley Close,
Weymouth, Dorset,
DT3 6DH, UK**

DAN DARE
PILOT OF THE FUTURE

The Anastasia, Tempus Frangit, Nimbus, Zylbat and associated Interplanet Space Fleet craft

Space Fleet Operations Manual

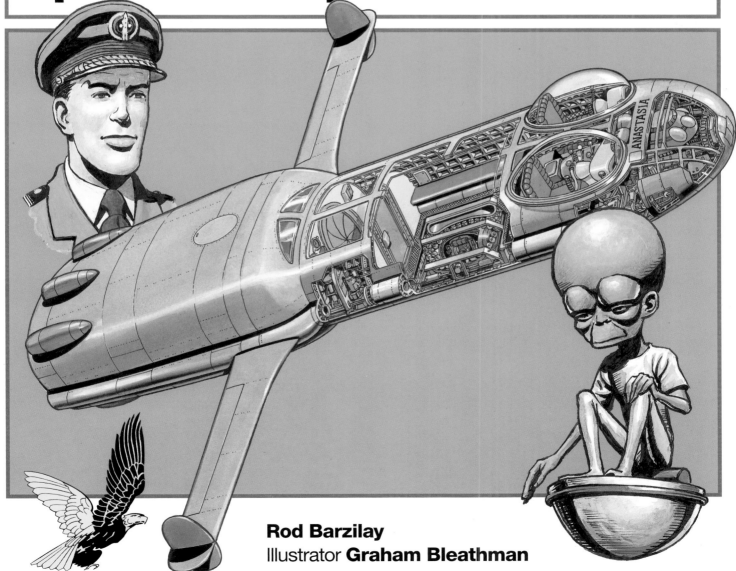

Rod Barzilay
Illustrator **Graham Bleathman**

CONTENTS

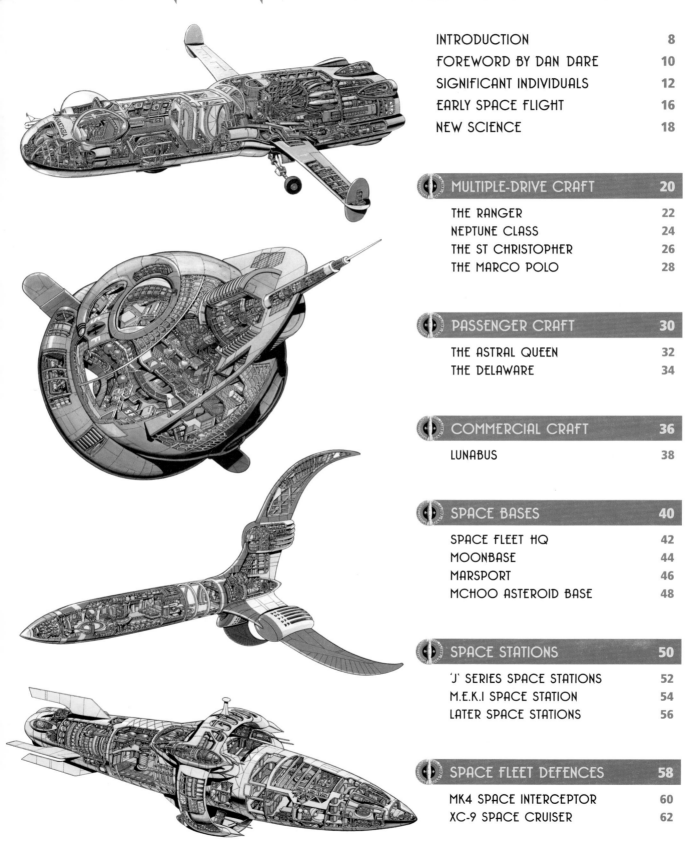

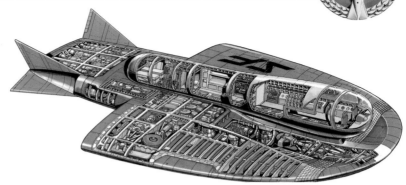
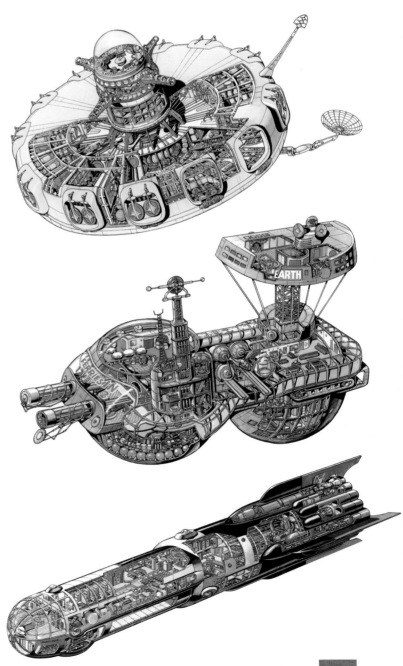

INTRODUCTION

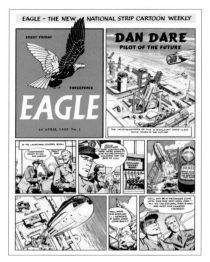

It was on 14 April 1950 that the innovative children's magazine EAGLE burst upon the publishing scene. The brainchild of parish vicar Marcus Morris and artist Frank Hampson, this publication had half its pages printed in full colour, and comprised a mixture of story strips and factual articles, including detailed vehicle cutaways. It took the world of children's publishing by storm. But the feature that really caught everyone's attention, and was to inspire a whole generation of future artists, writers and even scientists, was the Dan Dare – Pilot of the Future science fiction strip.

Based on science 'facts' of the time, with imaginative innovations thrown in for good measure, it was superbly drawn and scripted. Hampson and his team realistically portrayed future equipment and inventions, kindling the imagination of its readers and encouraging a widespread hankering for space travel and interplanetary adventure. At the time post-war Britain was full of optimism for a better and brighter future: the Festival of Britain was being prepared, the aviation industry was about to steal a world lead with its Comet aircraft, and the country was entering upon a boom age which would prompt prime minister Harold Macmillan to famously state 'our people have never had it so good'. So to many of us space travel felt just around the corner, and Hampson's exciting ideas didn't seem far-fetched at all.

The first Dan Dare tale was set 45 years into the future. It anticipated a world where Earth's population had doubled, and although science and, technology along with the UN, had made great strides, food production had failed to keep up, resulting in worldwide rationing (something that British youngsters were still experiencing in the 1950s, and could empathise with). Scientists in Dan Dare's time believed that food could be grown on Venus, thus solving Earth's food problems, and Space Fleet was trying to get spaceships there to see if this was possible. Thus the scene was set for Dan's first great adventure. Mars, Mercury, the moons of Saturn and even Triton would all feature in Dan's subsequent voyages, along with trips to other star systems.

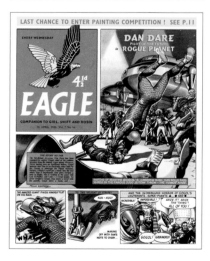

Some of Hampson's imagined inventions have now actually arrived, but others, sadly, have yet to materialise (oh for impulse engines and gravity motors!). Consequently, with a number of the strips' projected 1950s 'facts' having likewise failed the test of time, Dan's story has now become the future of an exciting parallel universe rather than our own. It nevertheless remains a fascinating place to be explored and investigated by readers and writers alike.

This book serves as a tribute to Hampson and his team, and all those who have made it such a rich place to visit. Come aboard and explore its diversity of people, spacecraft and astonishing technology.

Rod Barzilay
June 2013

Right: Frank Hampson burns the midnight oil working on his creation - with Dan Dare and the Mekon keeping an eye on things!

FOREWORD

E ver since the Venus expeditions in 1995 we have been exposed to repeated alien contact. Not only is our own solar system teeming with life – some hostile, some not – but various out-system life has been encounted, with mixed results.

Our expeditions into space have proved a steep learning curve for everyone, and we have not always got everything right. Getting it wrong has sometimes been very costly when alien technology has proved superior to our own, but we are doing our best to learn from our mistakes and to integrate other worlds' ideas and inventions into Space Fleet.

Space Fleet has been obliged to continually expand to meet these new challenges, and it is important not only to train our new recruits in as many alien technologies as possible, but also to keep everyone up to date regarding the impressive advances achieved by our own scientists.

However, preserving our knowledge of earlier technology has also proved invaluable – this is how, some years ago, two Space Fleet officers were able to keep themselves alive for nine years when stranded in a graveyard of lost spacecraft.

These are the reasons why I have commissioned this *Space Fleet Operations Manual*, which describes not only our own older and current craft but also important alien craft too, because when lost or alien spacecraft are encountered understanding their systems and capabilities may prove vital to our survival.

Dan Dare
Controller Space Fleet
February 2022

SIGNIFICANT INDIVIDUALS

Colonel*
Daniel MacGregor Dare

Space pioneer, participated in almost all of the important expedition teams from 1995 to 2019. A key figure in many of Space Fleet's finer moments, he has encountered and prevailed over the Mekon on a number of occasions, and solved many alien inter-planet issues. He is now the current controller of the Interplanet Space Fleet.

Professor*
Jocelyn Mable Peabody

Space pioneer and nutrition/agriculture/botany expert. Part of the Venus, Mercury, Saturn, Sargasso Sea of Space and Terra Nova expedition teams. Helped set up the Isle of Wight Interplanetary Zoo; came up with the idea of how to lure the Red Moon away from Earth; deciphered the Mercurian language. Her quick thinking saved lives on the journey to Terra Nova.

Class 1 Spaceman*
Albert Fitzwilliam Digby

Space pioneer and batman to Dan Dare since 1987, he has shared most missions with him, playing a key supporting role in many events and saving the day in some of them. Refuses any promotion that would prevent him from remaining Colonel Dare's batman.

Pilot Major*
Pierre August Lafayette

Space pioneer, part of the Venus, Asteroid and Saturn expedition teams. Realised that Dan Dare had hijacked a Telezero ship just in time to avert its destruction; assisted him during the Red Moon crisis.

Pilot Captain*
Henry Brennan 'Hank' Hogan

Space pioneer, part of the Venus, Asteroid and Saturn expedition teams. Made first contact with the Therons of southern Venus and helped pave the way to their alliance with Earth.

Marshal of Space*
Sir Hubert Gascoigne Guest

Space pioneer, a member of the first successful expeditions to the Moon, Mars, Venus and Terra Nova; succeeded Morgan Grosvenor as ISF Controller and founded the Astral Training College.

*The original Venus team. Ranks shown as at that time.

Sondar

The first Treen to understand Earth's way of thinking, and an important ally ever since. A spacecraft designer and pilot, led missions to the Red Moon and to Saturn. Designed Hermes spacecraft for Mercury, and Dan Dare's personal spaceship, the Anastasia. First off-world alien to befriend Tharl on Phobe. Became the first Treen ambassador to Earth, then governor and later president of northern Venus.

The Dapon in Chief

"The Dapon" was the regimental sergeant major of the first cohort of the Atlantine guards on Venus in 1996. He and his men were anti-Treen, and thinking that Dan was a reincarnated Atlantine hero named Kargaz they helped him escape from Mekonta. His final act was to cripple the Treens' feared Telezero weapon by suicidally crashing a spacecraft on to it, thereby eliminating the one weapon the Therons feared.

Anastasia Digby

Albert Digby's formidable aunt. It was she who recognised her nephew's coded message warning about the Treens' plans and ensured that Space Fleet understood the significance of it. It was through her efforts that the Mekon's 'Earth Plan' was frustrated. The following year Colonel Dare was presented with a specially built personal spaceship, designed by Sondar, that was named Anastasia in her honour.

Admiral Morgan Grosvenor

Space pioneer, leader of the first Mars expedition; prime mover in forming ISF, became its first Controller. When he died in 1989 his deputy, Sir Hubert Guest, succeeded him as Controller. A fleet maintenance and training ship has been named after him.

President Kalon

Leader of the Therons, persuaded by Dare to abandon his isolationist policy and ally with Earth against the Mekon. In 1999 he organised link with Treen scientists to try and find a way to control the Red Moon, but the attempt failed and the Red Moon was destroyed by a chain-reaction bomb.

Volstar

A Theron leader who assisted the original Venus expedition. He was President Kalon's emissary to Earth and later became leader of Theron–Atlantine resistance in the 2002–11 Robot War; secretly built the rocket bomb that destroyed the Selektrobot control satellite.

SIGNIFICANT INDIVIDUALS **13**

SIGNIFICANT INDIVIDUALS

Dr Blasco

A brilliant scientist who discovered a way to stabilise and use monatomic hydrogen as a rocket fuel. He designed the Valiant battlecraft for the run to Saturn, but had his own power agenda. Secretly in league with Rootha forces on Saturn, on his return he planned to take over the Earth. Foiled at the last moment, he lost his life when his space helmet came off in a struggle aboard the Kroopak command craft.

'Flamer' Toby Spry

An Astral Cadet involved in a number of Colonel Dare's adventures. When escaped prisoners Starbuck and Vulcani took over the Speedstar, he distracted them long enough for Colonel Dare to regain control of the ship. Also saved him from drowning on Cryptos. His most famous achievement was getting the Treens' Elektrobots to destroy themselves in 2011 by imitating the Mekon's voice.

Commander Lex O'Malley

Navy submariner and space pioneer. Set up the underwater limpet-grapple cable carriers for the recovery of Lero's spaceship from the Tuscarora deep; used his naval skills on both Cryptos and Terra Nova; located the Cosmobe and Pescod underwater bases; surveyed Earth's oceans after the planet's population had been evacuated to Mars; in charge of the Palk Strait Ocean Farm when the Fist organisation attacked.

Dr Ivor Dare

Colonel Dare's uncle. Archaeologist who translated the Martian writings, discovering much about the Red Moon. Accidentally part of the Sargasso Sea of Space expedition and stopped the Red Ship from self-destructing by working out the correct command sequence just in time.

Nikki

One-time Rootha radio operator who was a member of Tharl's rebel forces, and Dare's first Thork ally. Later became Tharl's Earth ambassador. Joined the Sargasso Sea of Space expeditions and got Tharl's flagship flying again. Was also able to decode Sakora's notes.

Tharl

Thork pirate and rebel leader who freed Titan from Rootha control. He also assisted Dan in rescuing his crew, and allowed the stripping of his fleet's radial-impulse generators to power the Valiant back to Earth. On his first interplanetary visit he was kidnapped with other VIPs by the Mekon.

Lero

Crypt emissary who came to Earth to recruit help against the Phants. The strength and capabilities of his spacecraft amazed Earth engineers. A number of concepts derived from it subsequently helped Space Fleet devise the Proto drive. When Lero left for Cryptos with the Earth team he passed on details of suspended animation techniques, first used by the Mekon, which would subsequently assist intergalactic travel.

Pilot Captain Bob 'Crusoe' King and Pilot Angus 'Friday' MacFarlane

Survived being stranded in the Sargasso Sea of Space for nine years. Helped Dan save the Anastasia spacecraft when it also became stranded there, then decoyed the Mekon's forces away from Dare's craft at a crucial time in 2011. Returned to the Sargasso Sea as guides with the Phoenix Mission the following year.

Katoona Kalon

Theron expert in many fields. Granddaughter of President Kalon, served with the Theron resistance during the 2002–11 Treen occupation. Joined the second expedition to the Sargasso Sea of Space, and made a number of discoveries. She also devised a way to release Tharl's spaceship from a gravity well while saving the lives of herself and six of Colonel Dare's team.

Galileo McHoo

Scientist, and chieftain of Clan McHoo, who rebuilt both his father's asteroid base and the Halley drive. Went on to design the Galactic Galleon that made mankind's first successful expedition to another star system and back, along with abducted ISF personnel.

'Primus'

Senior of three Ultimobe Cosmobes who commanded the Cosmobe spacecraft that detached from the Phantom Fleet to settle on Earth. Warned Earth of the hostile Pescod fleet that pursued them, and supplied a sample of 'Crimson Death'-proof material.

Colonel Wilf Banger

Led the team that investigated a time displacement spaceship found in the Sargasso Sea of Space, resulting in him creating the Tempus Frangit spacecraft. Served on its pioneering missions to Meit and Lapri. Was instrumental in destroying the Mekon's Mushroom structure in London.

EARLY SPACE FLIGHT

EARLY SPACE FLIGHT

Following the Second World War Britain took a great interest in outer space, experimenting and further developing captured German V2 rockets. In 1950 it launched the first Earth-orbiting satellite, Explorer 1.

The next challenge was to get a man into space, but this required a much more powerful rocket. It took five years to develop one capable of getting Commander Robert Findley into orbit in 1955 – another British first. By now America and France were also taking notice, and before long both had commenced their own space programmes. The USA's team was led by a very enthusiastic admiral, Morgan Grosvenor, who was determined that his country should not be left behind in the space race.

In 1956 the British rocket ship Janus was placed in a stationary orbit around Earth to serve as an emergency shelter and an in-space survival training centre.

1957 saw an unmanned spacecraft circumnavigate the Moon. A year later a two-man British craft piloted by Charles Wingate and John Breton succeeded in doing the same. Now, with the Americans catching up fast, the race was on to be the first country to land men on the Moon. Wingate, Breton and Bill Frazer attempted to achieve this in 1959, but disaster struck – the explosion of a fuel tank wrecked the spacecraft just after it entered Lunar orbit, killing Breton and Frazer and disabling the ship. Wingate survived by linking his spacesuit to the ship's own oxygen tanks until he was rescued by an American spacecraft commanded by Admiral Grosvenor.

Undeterred by this tragedy, the following year Wingate was in the British team that achieved the first successful Moon landing on 7 November 1960. Leader of the team was Andrew Harrison, who was the first man to step on to the surface of the Moon, followed by Hubert Gascoigne Guest and then Wingate.

Space travel was proving to be a very expensive adventure even for governments, and international ventures soon became the preferred way of financing expeditions. The first such enterprise saw Britain, America and France combining their resources to establish a fledgling Moon base between 1961 and 1963.

In 1965 an Anglo-American Mars expedition undertook an epic six-month voyage to the Red Planet in three ships, commanded by Grosvenor, Guest and Wingate. Landing on the Chryse Planitia, their crews spent a month exploring and surveying the surrounding area, collecting soil and rock samples. Then, as planned, the whole team returned to Earth in one craft. The other ships were left on Mars, and were later cannibalised for living quarters and storage tanks when a permanent base was established.

In 1966 the Interplanet Space Fleet was formed by Great Britain, the United States and France, under the auspices of the United Nations, with Admiral Grosvenor appointed as its Fleet Controller.

The ISF's first task was to build a major orbital satellite as its base. As this was before artificial gravity motors were invented, this took the form of a 'Space Wheel', which revolved in order to create centrifugal force and thus gravity on its outer edge. Ferry-craft docked next to the central weightless hub, where people and supplies transferred into an airlock and proceeded along the wheel's 'spokes' to the living and working quarters, where crews no longer needed to wear magnetic boots to stop them from floating away. This space station was created not only as a supply depot and emergency port but also as a construction facility where spacecraft and other equipment could be built in a gravity-free environment.

Right: Hermes III moon landing craft.

ISF DATA FILE

Along with space debris, including the *Janus* rocket, the Space Wheel was swept away when the Red Moon arrived. It was later rediscovered in the Sargasso Sea of Space.

MOON AND MARS LANDINGS

As already detailed, the first successful Moon landing took place on 7 November 1960. This was in the Sea of Tranquillity. Astonishingly at such a distance, this first expedition went almost entirely as planned. The only minor complication was that the ship overshot its intended landing spot because of a misjudged rocket burst.

After alighting on the surface, the three-man crew – Harrison, Guest and Wingate – planted a Union Flag and said a prayer of remembrance for John Breton and Bill Frazer, the two Englishmen who had died during the first attempt to land on the Moon, before collecting rock samples and taking photographs.

On a lighter note, they also tested who could jump the highest in the low lunar gravity. Wingate won with a spectacular leap and somersault. However, he landed awkwardly and it was sensibly decided not to push their luck any further. Discarding some of their ship's now empty rocket boosters to reduce weight and mass, the team took off and returned to Earth and a heroes' welcome.

Following the successful Anglo-American landing on Mars in 1965, in 1967 a larger mission was sent to the Red Planet to establish a permanent UN base. Amongst the extra equipment they took with them were sand sledges, allowing exploration teams access to a much larger area. This confirmed the long-suspected existence of artificial Martian canals and old ruins, which in turn would stimulate even greater interest in Mars back on Earth.

From 1969 Mars became the main focus of Space Fleet expeditions. Key questions were: could food be grown there (food shortages on Earth were now being predicted as a major problem for the future), and what could be learnt from the ancient Martian ruins? Most resources were therefore diverted towards the Red Planet. Although two probes had been sent towards Venus both had failed before reaching it, and the subsequent fascination with Mars meant that interest in Venus waned, and no more probes were sent.

ISF DATA FILE

The original landing site on the Moon, complete with footprints, discarded rocket boosters and flag, has now been domed over to form a museum, called 'Harrison's Footsteps'.

THE FIRST SHIPS TO MARS

Three spacecraft named after World War Two leaders made the first journey to Mars in 1965. One, the Charles de Gaulle, commanded by Wingate, was really a huge fuel supply ship. Before the invention of impulse power, spacecraft had to take all their fuel requirements with them; even though they coasted most of the way to Mars, it was important to maintain a sizeable fuel reserve in case of unexpected problems.

The Charles de Gaulle stayed in Mars orbit, ready for the return flight to Earth, but her crew transferred to Guest's ship, the Winston Churchill, which along with Grosvenor's Franklin Roosevelt landed on the Chryse Planitia. The Roosevelt landed horizontally, as the main base of the expedition, while the Churchill landed on its tail as a back-up base, its nose-section being a mini spacecraft that would ferry the entire team back to the Charles de Gaulle when it was time to leave. Likewise, the Roosevelt's nosecone could also act as a ferry if needed, but the plan was to leave the whole ship – along with the lower half of the Churchill – on the surface of Mars, for use by future expeditions.

The crews spent a month on the surface, exploring the local area around the ships, test-drilling and collecting soil, rock and air samples. Mars lift-off, and docking with the Charles de Gaulle, went as planned, and by using some of the fuel reserves to boost coasting speed the expedition got home a month sooner than expected.

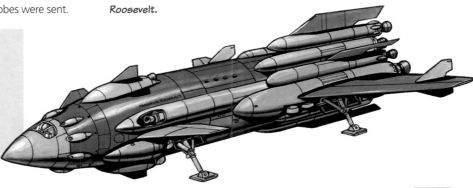

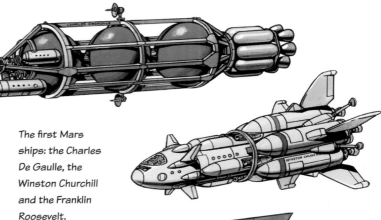

The first Mars ships: the Charles De Gaulle, the Winston Churchill and the Franklin Roosevelt.

IMPULSE DRIVE

In 1966 Alexis Hemmings announced the first Unified Theory of Physics, which modified Einstein's Theory of Relativity. Building on this he established the Second and Third Unifying Theories, now known as 'Hemmings' Theory', before going on to develop impulse drive and then gravity motors in the mid-1970s.

Today, in 2022, impulse waves are broadcast into space from ground stations on all the inner planets and around the asteroid belt. Spaceships pick up these waves and store them in impulse cylinders, like a battery stores electricity, from where the energy is fed to the engines as required; and as long as a ship is in an impulse field, or still has a reserve in its cylinders, it can continue to fly.

Impulse drive has two main advantages. Firstly, since spaceships no longer need to carry thousands of tons of fuel they can carry much more cargo. And secondly, spaceships can now boost their speed when required, instead of having to coast most of the time. This facility has dramatically reduced the time it takes to reach other planets and the inner Solar System.

Impressed with these advantages, Space Fleet adopted impulse drive in 1976 and rapidly started to expand its fleet of interplanetary ships. Then gravity motors arrived, making it much easier to work in space, and as a bonus eliminating the 'space motion sickness' many Space Fleet personnel experienced in weightless conditions.

In 1980 Edwin Hirshbaum devised a new, super-tough lightweight alloy, subsequently named 'Hirshinium', which resulted in lighter spacecraft that required far less power to boost them into orbit, which in turn led to even greater cost savings. With space travel now fast, comfortable and only a fraction of the old cost, commercial companies began to set up on both the Moon and Mars.

ISF DATA FILE

The invention of impulse drive led to the introduction of a new generation of spacecraft with multiple drive systems. Chemical rockets and solid fuel boosters still assisted in take-offs and emergency manoeuvres, but the main space drive was now by means of impulse waves. This led in turn to the introduction of faster, more economical spacecraft with much bigger payload capacities.

Space economics took a huge leap forward. Space travel now cost a tenth of what it had, which brought about rapid commercial and industrial development on the Moon and Mars.

Impulse Wave Generating Satellite Station
Powered by a nuclear reactor. There are 20 satellite generating stations spread throughout the solar system. All are automated but each contains life-support systems and accommodation for ISF personnel.

Asteroid-based Impulse Wave Generating Station
Built on to an asteroid, and located in the Asteroid Belt. Powered by a nuclear reactor, the station is automated, but does contain life-support systems and accommodation for maintenance and servicing personnel. There are 25 Generating Stations built on to asteroids, in addition to satellite stations and numerous Impulse Wave Relay transmitters attached to manned space stations such as the J-Series or commercial satellites.

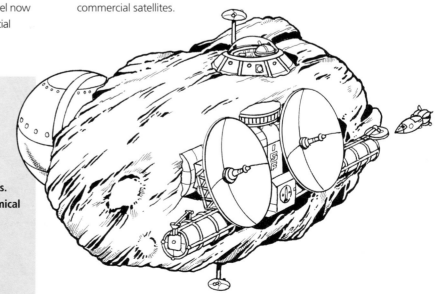

Ground-based Impulse Wave Generating Station

Although automated, Generating Stations are often manned to ensure maximum operational efficiency at all times. Eight are located on Earth, another six on the Moon and a further eight on Mars.

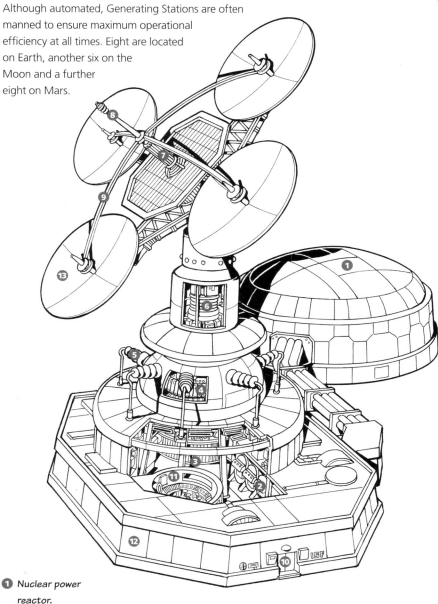

Gravity motors refocus gravitational forces into a new plane, or, where very little exists, create their own gravity field. In spacecraft, for example, gravity motors under the floors give a craft its own internal field, which will hold loose items to the floor in the same way that a planet's gravity holds things to its surface. The field can be set to a given strength, or even reversed so that things drift away or hover above the floor – which is very useful when moving large or heavy objects and for reaching ceilings without the need for ladders. Consequently gravity motors make it much easier to work in space.

On a planet, a re-focused gravity field can alter the direction of the gravitational plane within the localised vicinity of the motor, or if required the gravity plates themselves can be re-polarised by a gravity motor.

Gravity locks are double-doored cubicles that are built into the floor of each Launch Gantry embarkation corridor at SFHQ. If required, the lock is deployed onto the outer hatch of the spacecraft on the Launch ramp, and up to five personnel can enter at once. The floor rotates up to 90 degrees, adjusting the gravity direction to match that of the spacecraft.

Below: A gravity lock in use.

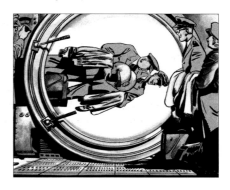

1. Nuclear power reactor.
2. Electricity generating turbines.
3. Impulse wave power amplifier.
4. Transmitter sub-system generator.
5. Transmitter generator cooling coils.
6. Signal strength accumulator coils.
7. Primary beam focusing generator.
8. Focused impulse beam emitter.
9. Wide beam distribution conduit.
10. Guarded entrance.
11. Control Room.
12. Accommodation and offices.
13. Wide beam impulse wave emitter also transmits and receives audio, video and data communication to space fleet vessels faster than standard radio signals.

MULTIPLE-DRIVE CRAFT

VENTURE-CLASS SHIPS

These were devised to service the Moon and Mars surface bases, and commenced operation in 1983. For years they were the heavy-duty workhorses of the ISF fleet. Large and sturdily built spacecraft with 12-man crews, they served Space Fleet well until 2001, when the last of them were replaced by Neptune-Class ships, which needed crews of just six men.

ISF DATA FILE

It was Venture-Class craft that made the original attempts to reach Venus. Sadly three were lost, the Orion and Argonaut in 1994, and the Kingfisher in 1995. Only the Ranger survived, because it didn't approach any closer than 3,000 miles from the planet.

Above: The Kingfisher on its way to Venus.

NEPTUNE-CLASS SHIPS

These ships were originally devised for the food runs between Venus and Earth, but, like the Venture-Class ships they replaced, they became invaluable workhorses. They were utilised to transport large loads into space for the construction of the ISF's space stations, and the space trains that use them, as well as the impulse relay stations dotted around the asteroid belt. Compared to the Venture class, they were taller and slimmer, could carry more cargo and were operated by smaller crews of just six (three on duty, three resting); with autopilot aid, miners often crewed them with even fewer, and in some circumstances they could be operated with a crew of two. As a testament to their reliability, even now – 25 years after the first Neptune went into service – some can still be found ferrying supplies to the miners of the asteroid belt and ore back to Mars. The Neptune class may be an old design, but it remains a worthy one.

Above: Neptune–Class rockets take off to bomb the Red Moon.

ISF DATA FILE

Three Neptune-Class spaceships were adapted to carry atomic bomb missiles in an attempt to stop the Red Moon from reaching Earth in 1999. All 12 bombs hit the target areas, but they failed to stop the Red Moon.

SPACE FERRIES

These tough, versatile little ships with a two-man crew carry up to 14 passengers and their baggage at a time, from a planet's surface to a satellite space station or back. They can take off and land on runways, or boost into space from ramps, and are slim enough to enter a space station's docking tubes once the ship's wings are retracted. They have an upgraded impulse engine, supplemented by two rocket boosters.

Left and above: Two views of a Mars ferry.

RESCUE SHIPS

The loss of the American spaceship Caroline with her entire crew in 1978 was the catalyst that brought about the design and introduction of space rescue ships. The Caroline suffered engine failure just as she reached orbital velocity, but being at too low an orbit air-drag brought on orbital decay, and she started to slowly fall back towards Earth. By the time a second spaceship could be prepared, crewed and sent to the area it was only in time to witness the *Caroline* burning up in Earth's atmosphere. If the relief craft had got there just a little sooner everyone could have been trans-shipped to safety.

After this spaceports around the world had prepped rescue ships on permanent standby. In an emergency a space alarm would sound, and as all space crew were taught first aid and basic engineering the first crew to reach the rescue ship would take her up. If necessary the rescue team could radio their base for further advice or assistance.

ISF DATA FILE

The strangest mission that a rescue ship has yet flown occurred when a tranquilised Venusian triceratops en route to an Earth zoo woke up and smashed the control runs to the steering jets.

Right: A rescue ship in action.

THE MARCO POLO EXPLORATION CRAFT

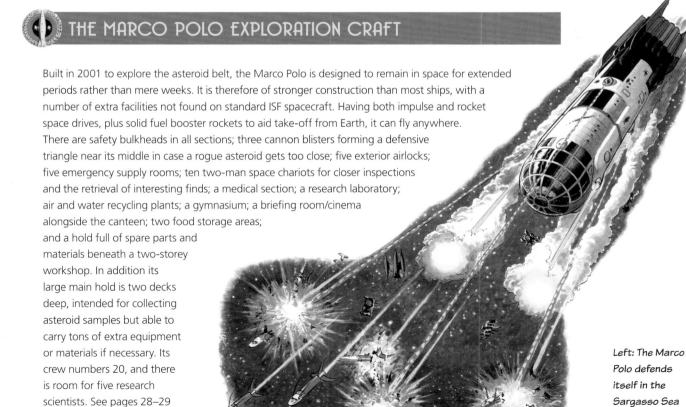

Built in 2001 to explore the asteroid belt, the Marco Polo is designed to remain in space for extended periods rather than mere weeks. It is therefore of stronger construction than most ships, with a number of extra facilities not found on standard ISF spacecraft. Having both impulse and rocket space drives, plus solid fuel booster rockets to aid take-off from Earth, it can fly anywhere. There are safety bulkheads in all sections; three cannon blisters forming a defensive triangle near its middle in case a rogue asteroid gets too close; five exterior airlocks; five emergency supply rooms; ten two-man space chariots for closer inspections and the retrieval of interesting finds; a medical section; a research laboratory; air and water recycling plants; a gymnasium; a briefing room/cinema alongside the canteen; two food storage areas; and a hold full of spare parts and materials beneath a two-storey workshop. In addition its large main hold is two decks deep, intended for collecting asteroid samples but able to carry tons of extra equipment or materials if necessary. Its crew numbers 20, and there is room for five research scientists. See pages 28–29 for a detailed cutaway.

Left: The Marco Polo defends itself in the Sargasso Sea of Space.

Because of its involvement with the first successful Venus expedition, the Ranger is the most famous spacecraft of the Venture class. It carried three small rocket ships as far as the ray field that surrounded the planet, one attached to each side of the Ranger's hull while the third was stowed in her hold. These rocket ships were launched 3,000 miles out from the planet. They made it through the ray field – although one was damaged in the process – and then flew on towards Venus and into the history books.

Right: Rocket 3 deploying.

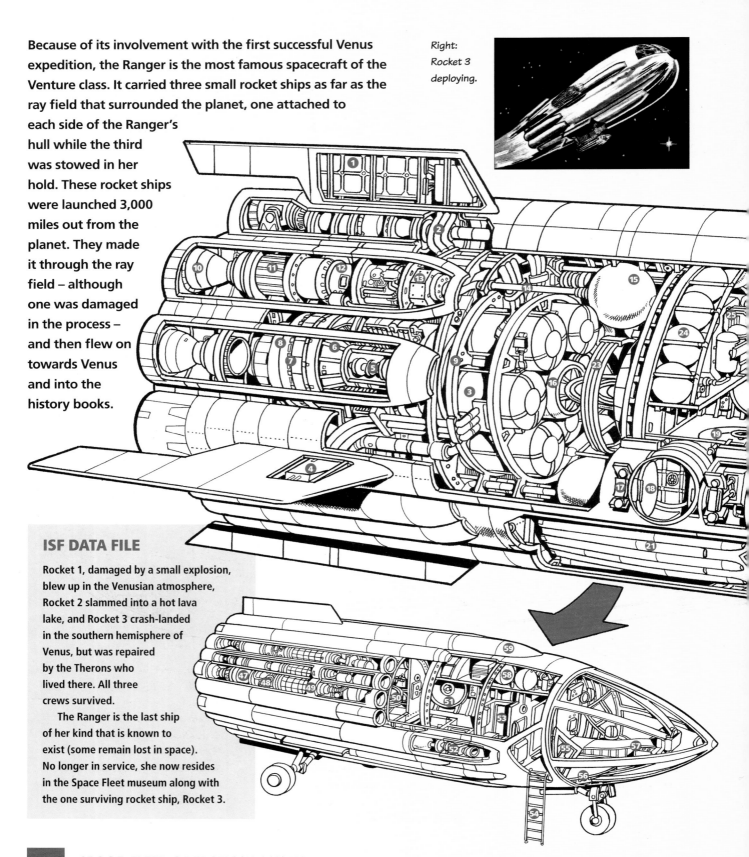

ISF DATA FILE

Rocket 1, damaged by a small explosion, blew up in the Venusian atmosphere, Rocket 2 slammed into a hot lava lake, and Rocket 3 crash-landed in the southern hemisphere of Venus, but was repaired by the Therons who lived there. All three crews survived.

The Ranger is the last ship of her kind that is known to exist (some remain lost in space). No longer in service, she now resides in the Space Fleet museum along with the one surviving rocket ship, Rocket 3.

1. Impulse wave receptor panels.
2. Impulse wave power conduits.
3. Impulse wave accumulator/propellant cylinders.
4. Impulse wave energy converter.
5. Particle accelerator.
6. Ionisation chamber.
7. Ionisation grid ring attachments.
8. Ion thrust nozzle.
9. Electro-sphere.

19. Airlocked access hatch to rocketship housed in underside hangar.
20. Airlock to portside rocket ship.
21. Rocket ship housed in underside hangar.
22. Hangar bay doors.
23. Gyroscope controls ship's pitch and yaw in conjunction with retro rockets.
24. Compressed air tanks.
25. Air scrubbing and recycling systems.
26. Rear crew quarters.
27. Topside retro rocket.
28. Starboard airlock outer hatch.

38. Artificial gravity generator.
39. Gimballed starboard retro rocket.
40. Life-support systems consoles.
41. Access to avionics, gravity control systems and underside rocketship hangar.
42. Engineering and flight deck.
43. Polarised, heat and radiation resistant flight deck canopy.
44. Pilot's couch.
45. Primary flight controls.
46. Avionics systems bay.
47. Chemical rocket combustion chamber.
48. Oxidant tank.
49. Chemical rocket fuel tanks.
50. Gyroscope assembly.

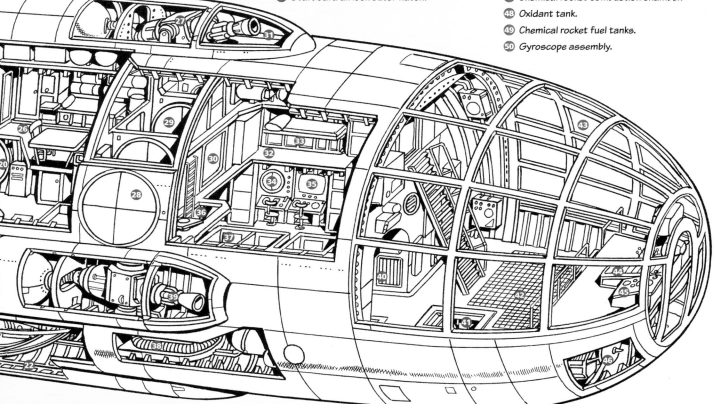

10. Chemical rocket thrust nozzle.
11. Combustion chamber.
12. Turbine and pumps.
13. Oxidant tanks.
14. Fuel feed pipe.
15. Chemical rocket propellant tanks.
16. Electricity generator.
17. Localised gravity field clamp connects scoutship to starboard side of hull.
18. Outer airlock access door to starboard rocketship.

29. Port airlock inner hatch.
30. Access door to rear crew's quarters and main airlocks.
31. Retro rocket cowling hatch.
32. Forward crew's quarters.
33. Sleeping bunk and personal storage.
34. Navigation station.
35. Communications console and video screen.
36. Access hatch connecting crews' quarters to flight deck.
37. Space suit storage.

51. Portside airlock door connecting Venus rocketship to Ranger's starboard flank.
52. Starboard retro rocket.
53. Personal hygiene station.
54. Access ladder to flight deck.
55. Flight deck access hatch.
56. Nosewheel stowage bay under flight deck.
57. Pilot's couch and control console.
58. Compressed air tanks.
59. Retro rocket protective cowling hatch.

The Neptune Class became the favoured workhorse of the ISF fleet after their introduction in late 1996, and they were successfully used for many duties, including aiding in the construction of space stations and the transportation of equipment and food between planets. Three were even converted into a triple atomic bomb-carrying ship (like the cutaway shown below) when the Red Moon was heading for Earth.

Above: Neptune-Class ship descending on landing jets and rotor blades.

1. Toughened laminate polarised cabin window, bonded with ceramic additives to render it heat resistant. Also radiation and solar glare resistant.
2. Radar scanner.
3. Avionics bay.
4. Pilot's couch.
5. Control room.
6. Navigation station.
7. Port air lock and access ladder.
8. Forward port VTOL rocket engine.
9. Communications station.
10. Forward landing leg.
11. Airtight topside rotor assembly doors.

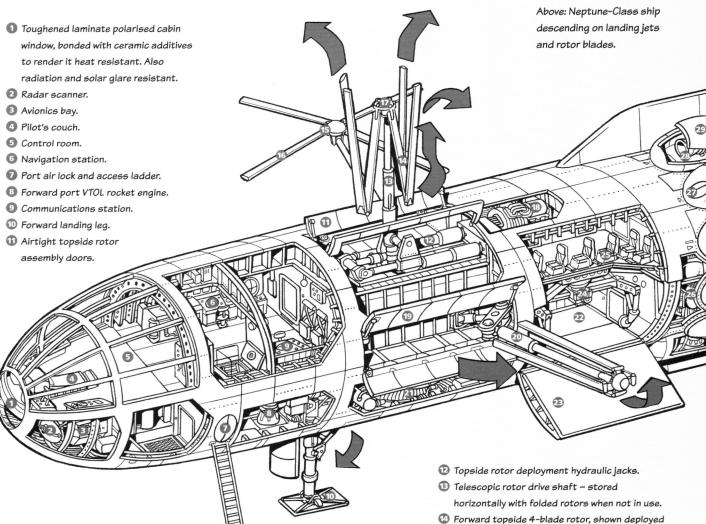

12. Topside rotor deployment hydraulic jacks.
13. Telescopic rotor drive shaft – stored horizontally with folded rotors when not in use.
14. Forward topside 4-blade rotor, shown deployed but in folded position.

15 Blade root attachment joint.

16 Forward starboard 4-blade rotor assembly.

17 Rotor head hydraulic reservoir.

18 Topside rotor assembly electric motor

19 Port rotor assembly air tight doors.

20 Forward port 4-blade assembly shown deployed but in folded position.

21 Artificial gravity generator provides power for gravity plating on all decks.

22 Underside cargo bay doors used as storage deck if side doors only are used.

30 Retro rocket.

31 Gimballed pitch and yaw attitude correction rocket.

32 Topside rotor blade deployment jack.

33 Air tight topside rotor assembly doors.

34 Retro rocket fuel tanks.

35 Rear topside 4-blade rotor assembly with folding blades shown fully deployed.

36 Topside rotor assembly electric motor.

42 Port rotor assembly air tight doors.

43 Rear port VTOL rocket engine.

44 Rear port landing leg.

45 Landing leg hydraulic jacks.

46 Rear port landing leg door.

47 Rear port 4-blade rotor assembly shown in deployed position.

48 Ion stream exhaust nozzle.

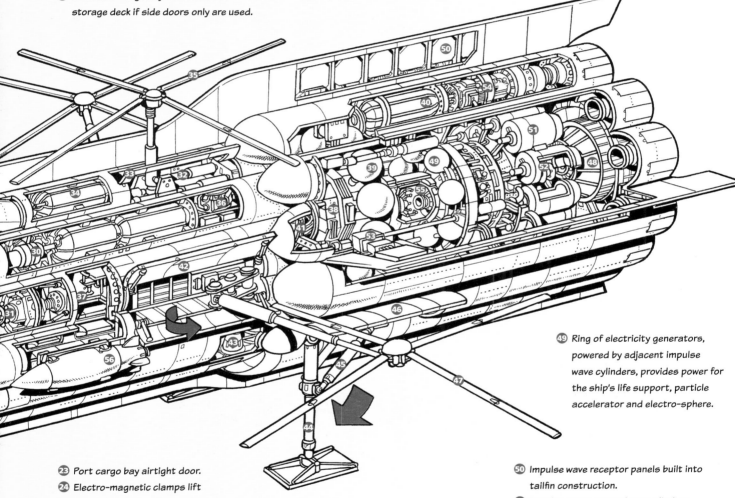

23 Port cargo bay airtight door.

24 Electro-magnetic clamps lift cargo from either port, starboard or underside access doors.

25 Additional personnel/crew lift-off and acceleration couches.

26 Midships port VTOL rocket engine.

27 Retro rocket nozzle gate seal.

28 Gimballed pitch and yaw rocket.

29 Hinged pitch and yaw rocket cowling hatch.

37 Crew quarters.

38 Air recycling and life-support systems.

39 Compressed air tanks.

40 Rocket fuel tank.

41 Attitude correction gyroscope, used in conjunction with pitch and yaw rockets.

49 Ring of electricity generators, powered by adjacent impulse wave cylinders, provides power for the ship's life support, particle accelerator and electro-sphere.

50 Impulse wave receptor panels built into tailfin construction.

51 Impulse wave accumulator cylinders.

52 Particle accelerator.

53 Impulse wave power converter.

54 Take-off booster rocket used if lifting off vertically from launch pad.

55 Ion accelerator electro-sphere.

56 Atomic bomb: used – unsuccessfully – to stop the Red Moon reaching earth.

THE ST CHRISTOPHER

The St Christopher is ISF HQ's second rescue ship, the first having suffered so much damage during the Red Moon attack in 1999 that it had to be replaced. The St Christopher is equipped with powerful boosters that enable it to reach distressed spacecraft as quickly as possible, and it carries a variety of life-support equipment and emergency repair gear. Designed for a flight crew of three plus two specialists, it can carry more if needed.

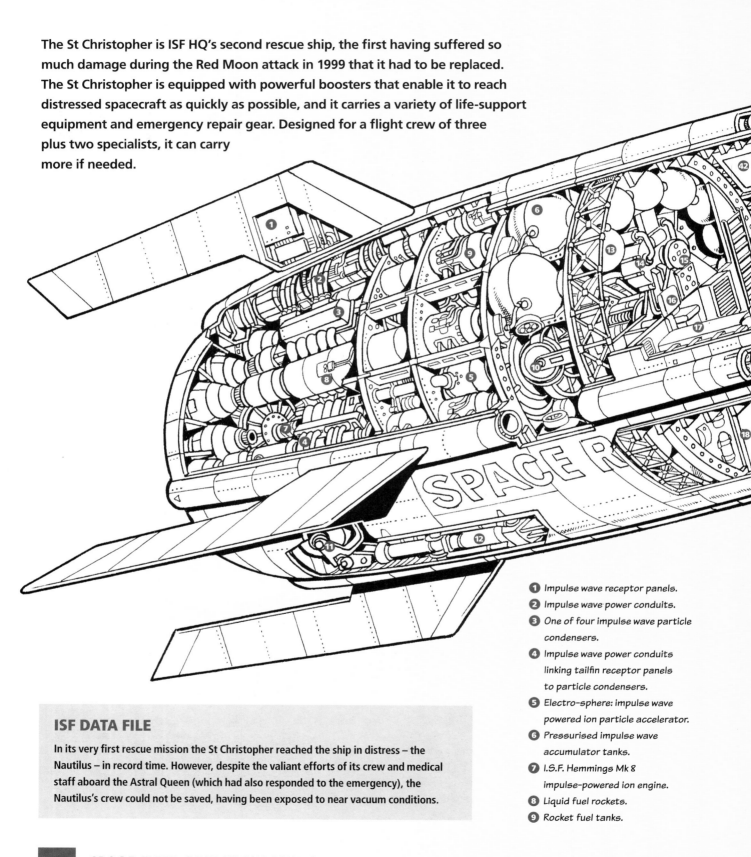

ISF DATA FILE

In its very first rescue mission the St Christopher reached the ship in distress – the Nautilus – in record time. However, despite the valiant efforts of its crew and medical staff aboard the Astral Queen (which had also responded to the emergency), the Nautilus's crew could not be saved, having been exposed to near vacuum conditions.

1 Impulse wave receptor panels.
2 Impulse wave power conduits.
3 One of four impulse wave particle condensers.
4 Impulse wave power conduits linking tailfin receptor panels to particle condensers.
5 Electro-sphere: impulse wave powered ion particle accelerator.
6 Pressurised impulse wave accumulator tanks.
7 I.S.F. Hemmings Mk 8 impulse-powered ion engine.
8 Liquid fuel rockets.
9 Rocket fuel tanks.

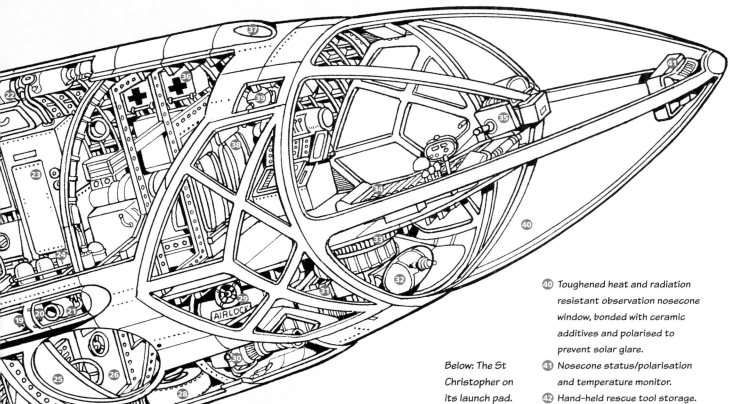

40 Toughened heat and radiation
resistant observation nosecone
window, bonded with ceramic
additives and polarised to
prevent solar glare.

Below: The St
Christopher on
its launch pad.

41 Nosecone status/polarisation
and temperature monitor.

42 Hand-held rescue tool storage.

10 Attitude correction gyroscope
used in conjunction with retro
rockets and impulse engines to
alter course during spaceflight.

11 Starboard mainwheel.

12 Starboard landing wheel strut
(retracted).

13 Compressed air tanks.

14 Air-recycling and life-support
systems.

15 Life-support controls.

16 Medical officers' seat.

17 Medical bay patients bed.

18 Space suit storage lockers.

19 Attitude rocket fuel tank.

20 Attitude rocket protection hatch
remains closed during re-entry
or when rocket is not in use.

21 Attitude gimballed control rocket.

22 Water tanks.

23 Personal hygiene station.

24 Emergency oxygen cylinders.

25 Starboard outer air lock hatch.

26 Inner airlock door leading to
lower deck.

27 Nose wheel doors.

28 Nose wheel rotated 90
degrees in stored position.

29 Port outer airlock hatch.

30 Underside gimballed retro/attitude
correction rocket.

31 Avionics bay.

32 Forward radar and sensor array.

33 Gravity deck plate power generator.

34 Pilot's couch.

35 Video screen and flight controls.

36 Medical equipment storage lockers.

37 Long range camera.

38 Port bulkhead-mounted crew seats.

39 Hand operated view-screen
linked to camera above and
forward scanners beneath
the cockpit deck.

The Explorer-Class Marco Polo represents the pinnacle of ISF engineering achievement. Thanks to the flexibility of its multi-drive propulsion systems, the craft can operate comfortably both within and outside the range of Impulse Wave fields for long-duration missions. The Marco Polo shot to fame in 2012 on a mission to recover the Z-19 spacecraft from the Sargasso Sea of Space.

Keith Watson

Above: The Marco Polo's central flight deck.

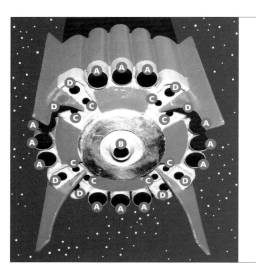

MULTI DRIVE SYSTEM

A Solid fuel rocket boosters x12. Used to boost ship at take-off only.

B Main engine – Impulse wave. Only used when in impulse wave fields. Not in operation while in the Sargasso Sea of Space.

C Impulse wave steering rockets x8. Again, not used in the Sargasso Sea of Space.

D Liquid fuel rockets x8. Only propulsion used in the Sargasso Sea of Space (apart from front retro-rockets). Different combinations can be used to steer the ship, all firing at once for straight-line flight.

1 Polarised observation windows.

2 Flight controls.

3 Acceleration couch.

4 Pilot's couch.

5 Lower observation deck.

6 Periscope monitor.

7 Hatch to forward hold.

8 Flight deck.

9 Forward storage hold.

10 Retro-rocket fuel tank.

11 Observation dome, linked to periscope.

MARCO POLO

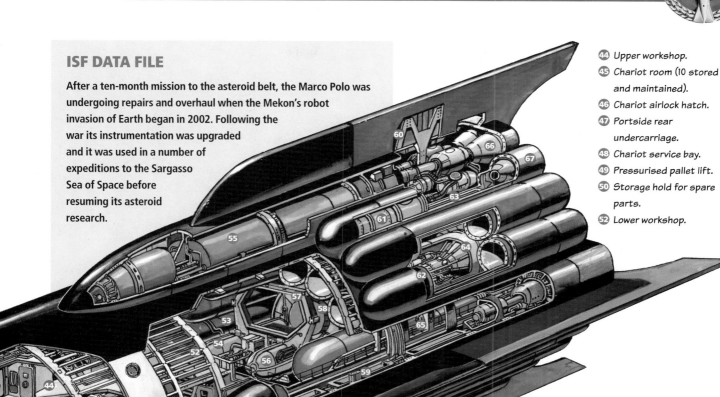

ISF DATA FILE

After a ten-month mission to the asteroid belt, the Marco Polo was undergoing repairs and overhaul when the Mekon's robot invasion of Earth began in 2002. Following the war its instrumentation was upgraded and it was used in a number of expeditions to the Sargasso Sea of Space before resuming its asteroid research.

44 Upper workshop.
45 Chariot room (10 stored and maintained).
46 Chariot airlock hatch.
47 Portside rear undercarriage.
48 Chariot service bay.
49 Pressurised pallet lift.
50 Storage hold for spare parts.
52 Lower workshop.

12 Upper deck central corridor bulkhead door.
13 Toilets.
14 Upper observation deck.
15 Upper deck storage.
16 Life-support conduits.
17 Ship commander's cabin.
18 Retro-rocket engine.
19 Swivel-mounted, multi-directional rockets (port and starboard).
20 Portside forward undercarriage.
21 Forward portside crew cabins (x3).
22 Main airlock with telescopic transfer tube.
23 Spacesuit lockers and tools.
24 Forward starboard crew cabins (x5).
25 Central upper deck corridor.

26 Gymnasium.
27 Air and water recycling systems.
28 Compressed air tanks.
29 Main cargo hold.
30 Skittle alley.
31 Crew mess.
32 Doctor's surgery.
33 Food preparation area.
34 Bar and food serving area
35 Briefing room/cinema (seats 25).
36 Laboratory.
37 Cold store.
38 Quartermaster's office.
39 Laundry.
40 Disintegrator defence cannon turrets (upper, port and starboard).
41 Officer's/VIP cabins.
42 Portside chariot room airlock.
43 Storage area.

51 Herculeneum strengthened structure contains air pumps and additional life-support systems.
53 One of 12 rocket fuel tanks.
54 Particle impulse wave recifier.
55 Liquid fuel tanks.
56 Impulse wave particle condenser.
57 Forward internal fuel tank support stanchion.
58 Electro-sphere: primary ion particle impulse accelerator.
59 Liquid fuel tank.
60 Herculeneum strengthened stabilising fin.
61 Pressurised condensed solid fuel propellant tanks.
62 Secondary impulse wave particle accelerator.
63 Impulse wave steering rockets.
64 Rocket combustion chamber for main impulse wave engine (used when in impulse wave fields).
65 Maintenance gantry.
66 Liquid fuel rockets used for general propulsion and steering.
67 Solid fuel rocket boosters used to achieve escape velocity during take-off.

PASSENGER CRAFT

SPACE FLEET PASSENGER CRAFT

Space Fleet was created to explore the solar system and search for anything that might ultimately benefit mankind. From the very beginning, Mars in particular attracted huge interest amongst scientists seeking ways to grow more food to solve the problem of food shortages on Earth, and when the ruins of a past civilisation were found archaeologists also became interested. The introduction of impulse power, and with it cheaper space travel, led to commercial development and the start of interplanetary tourism. Mars became a popular holiday destination, not only with people coming to see the Martian ruins, but to use the new Ski resorts at the North Pole. Space Fleet, seeing this as a way to recover some of its operating costs, consequently added passenger accommodation to a number of its ships.

In 1997 Space Fleet built a new passenger flagship for the Mars run. This was the Astral

Queen, capable of carrying a hundred passengers and crew. She became a favourite with travellers and remained so even after the introduction of space trains the following year. This was because she could go direct from planet to planet, without any need to change ship at the each end of the journey, as is necessary when using space trains. However, her popularity eventually waned following the arrival of luxury space trains in 2014.

Space trains were a logical development of space travel. Built to operate only in the vacuum of space, between space stations, they have no need for streamlining since they aren't required to descend or ascend through an atmosphere, or cope with the tremendous gravitational stresses of take-off. They are even assembled in space.

Most space trains follow the same basic design – two linked globes riding on a large, un-pressurised cargo hold. For extra safety the

Above: The Lancastrian arrives at SFJ2 to find Operation Lifeboat in full swing. The drive unit can be rotated above the spacecraft to act as a brake.

propulsion pod is in a separate section attached to the craft by two movable arms, which can not only move the impulse-powered cluster above or below the rest of the train, but can also swing it to either side along a curved track in order to steer the craft. The drive section can also rotate on the arms above the ship to slow and halt its forward momentum.

The first space trains – such as the Space Clipper, Maryland, Lancastrian and Magenta – had much bigger cargo holds than they had passenger-carrying capacity, but as space travel became more popular some craft switched their priorities. Space trains can now be found on both the Earth–Mars and Earth–Venus runs.

Above: The Astral Queen.

Right: The Maryland space train.

ISF DATA FILE

The Space Clipper, Maryland and Lancastrian were all involved in the emergency evacuation of Mars in 1999, when the Red Mood struck.

LUXURY SPACE TRAINS

As has been mentioned, the early space trains favoured cargo space over people-carrying capacity, but the builders of the Delaware, launched in 2002, put the emphasis on its passengers, focusing more on comfort and novelty in order to attract premium customers. Its facilities included a gravity-free padded play dome with panoramic views of space; a spectacular two-deck 170° wrap-round cinema; a full ship-length shopping mall; and a star-view bar and dance floor, galaxy restaurant, coffee lounge and games room – all with sensational space views. In addition it carried three life-craft as part of its well-advertised safety features.

The drive cluster was modified too. It could still move up and down on its supporting arms, and roll 180° above the ship to cancel out forward motion, but gone was the circular track that standard space trains used to position their thrusters for steering. Instead some of the boost tubes were set at an angle to allow all-round pitch and yaw control.

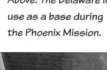

Above: The Delaware in use as a base during the Phoenix Mission.

Right: The Delaware is struck by debris during the Phoenix Mission.

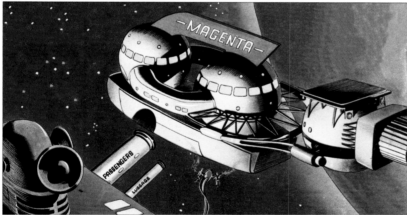

Above: The Magenta undocking from Space Station Ten in Earth orbit.

Right: The floodlit Space Clipper moored at SFJ2 while in the shadow of Mars.

ISF DATA FILE

The Delaware's three life boats were put to good use on the return leg of its maiden voyage to Mars in January 2002 when the drive pod was crippled by a meteor strike. As the craft was drifting out of impulse-power range, passengers and crew were obliged to abandon ship. An operation to recover the *Delaware* was planned, but the Mekon's robot invasion of Earth intervened, and it drifted on until it was swept up into the Sargasso Sea of Space. Here it was of great help to two Space Fleet officers whose own crippled ship had been pulled into the area, keeping them alive for nine years. In 2012 some of the Phoenix Mission team used it as a base, until it sustained further damage from flying wreckage following an explosion. It was eventually salvaged in 2014 and has now returned to the Mars run.

Space Fleet designed the Astral Queen as a passenger-carrying flagship for the Mars run. She remained very popular for some years, but by 2014 was losing ground to newer luxury craft, and when passenger numbers dropped too low for her to be run profitably as a liner she was converted into a cargo ship. All her passenger berths and lifeboat bays were then transformed into extra freight holds.

Above: Lifeboats leaving the Astral Queen.

1 Toughened laminate polarised forward observation window. Bonded with ceramic additives, the window is both radiation and heat resistant, impervious to high atmospheric re-entry temperatures.

2 Observation lounge bar.

3 Forward observation lounge.

4 Computer and instrument panel shroud.

5 Control room.

6 Pilot and co-pilot's seats.

7 Long-range scanner/camera operator's position.

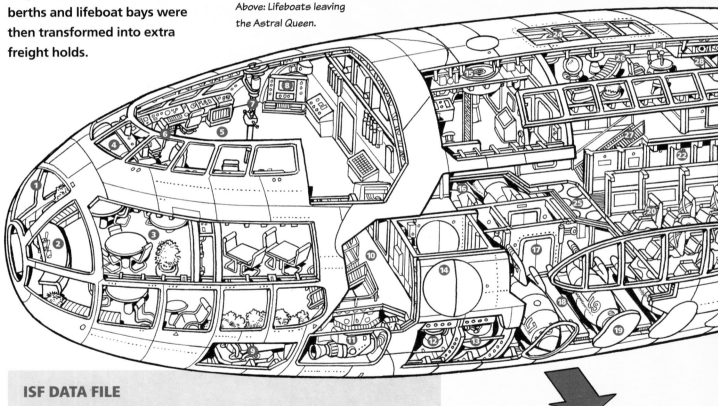

ISF DATA FILE

The Astral Queen was involved in two space emergencies, the first in 1997 when she was used for the dramatic rescue of a Space Fleet survey ship trapped in the asteroid belt. The official inquiry into this event commented that it was a pity the Astral Queen didn't have better medical facilities to treat injured spacemen. To counter this shortcoming Space Fleet converted one of the ship's recreational areas into a hospital ward. Consequently when, in 2000, a 'Black Cat' attack depressurised the Nautilus, the Astral Queen was called in by rescue craft to try and save the crew, but unfortunately without success.

8 Navigation station.

9 Forward port telescopic landing leg (retracted).

10 Observation lounge balcony.

11 Forward port gimballed retro/pitch and yaw rocket.

12 Forward port VTOL thruster.

13 VTOL thruster fuel tank.

14 Port airlock outer hatch; provides primary embarkation access for passengers.

15 Topside crew airlock inner hatch.

16 Starboard escape pods.

17 Port forward escape pod deck access door.

18 Port side escape pod deck.

19 Escape pod hatches.

20 Acceleration couches: used by passengers when the Astral Queen takes off from an upright position. Seating can be reconfigured to provide comfortable viewing of films (3D screen not shown).

33 Altitude correction gyro-scope.

34 Ion accelerator electro-sphere.

35 Impulse wave power converter.

36 Impulse engine support bulkhead.

37 Engine maintenance access chute.

38 Ion particle gun.

39 Impulse wave accumulation cylinders.

40 Ion particle accelerator.

41 Impulse wave receptor panels built into tailfin construction.

42 Upper tailfin launch support stanchion.

43 One of four tailfin mounted booster rockets.

44 Rear hull mounted impulse-wave receptors.

45 Ion particle outlet nozzle.

46 Tailfin lift-off booster rocket cowling.

47 Booster rocket linkage wells.

48 External booster rocket.

ESCAPE POD

A Forward parachute deployment hatch.

B Life-support systems.

C Forward landing leg (deployed).

D Fire extinguisher.

E Topside landing parachute storage and deployment mechanism.

F Landing leg (deployed).

G Fuel tanks provide limited distance travel or controlled planetary landfall.

H Tailfin-mounted rocket engines.

I Forward facing seats

J Rear-facing seat.

21 Gravity generator.

22 Lower passenger deck central corridor.

23 Access door to starboard acceleration couch deck.

24 Topside lounge observation windows; polarised, heat and radiation resistant.

25 Artificial gravity plating on all decks.

26 Topside passenger lounge.

27 Stairwell and adjacent lift to all decks.

28 Bar.

29 Passenger cabin deck central corridor.

30 Access to washroom and toilets.

31 Emergency space suit lockers.

32 Atmosphere recycling and life-support systems.

49 Booster rocket fuel tanks.

50 External take-off booster rocket engine.

51 External underside booster rocket electro-magnetic linkage.

52 Port tailfin leading edge strengthening stanchion supports the weight of the ship in upright position prior to launch.

53 Rear port telescopic landing leg (retracted).

54 External booster rocket: one of four connected to each tailfin supplying additional power during initial take-off. When the fuel is spent, each rocket detaches and returns to the ground by parachute.

THE DELAWARE

Once the Delaware was back in service in 2014, after being salvaged from the Sargasso Sea of Space and refurbished, she became a top attraction that – despite being a space train – even enticed passengers away from the Astral Queen.

Above: Onboard posters from the Delaware.

1. Pilot's console.
2. Captain's position.
3. Flight deck.
4. Navigation station.
5. Communications station.
6. Engineering station.
7. Stairs to all decks.
8. Inter-deck lift.
9. Star view bar lounge.
10. Dance floor.
11. Galaxy restaurant.
12. Coffee lounge.

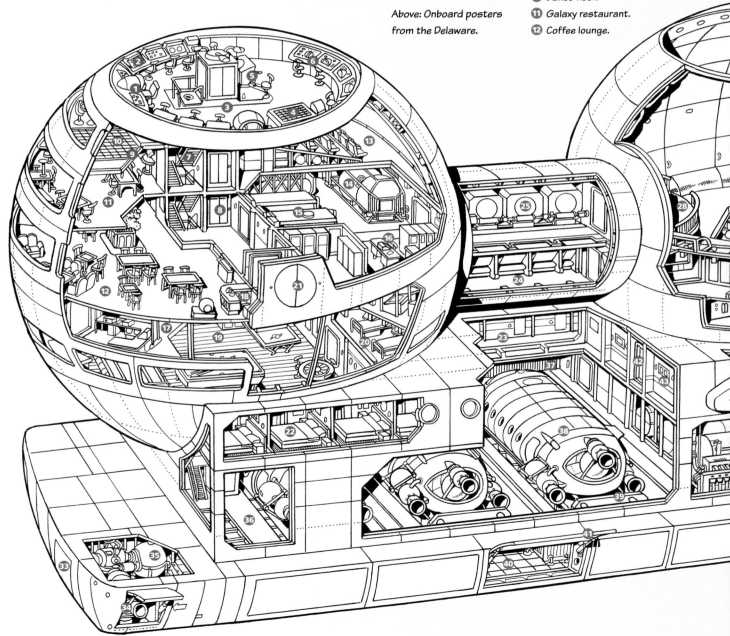

13 *Spectacularama 170º cinema.*
14 *Air recycling tanks.*
15 *Water purification and filtration tanks.*
16 *Environment control section.*
17 *In-flight shopping deck.*
18 *Coffee bar.*
19 *Games room.*
20 *Medical bay.*

21 *Passenger airlock.*
22 *Passenger sleeping cabins with starview window.*
23 *Starboard passenger deck corridor.*
24 *Access corridor linking forward and rear shopping decks.*
25 *Compressed air tanks.*
26 *Gravity-free play dome with padded walls to prevent injury.*
27 *Spaceview topside dome, polarised against solar glare and radiation.*
28 *Play dome balcony.*

29 *Play dome passenger protection padding.*
30 *Artificial gravity generator powers deck plating motors and also reduces, varies or eliminates gravity in the play dome above.*
31 *Rear in-flight shopping deck.*
32 *Crew quarters.*
33 *Port forward manoeuvring rocket.*

34 *Port retro/manoeuvring/ docking rocket.*
35 *Manoeuvring rocket fuel tank.*
36 *Lifeboat bay passenger assembly point.*
37 *Lifeboat bay door.*
38 *Passenger escape lifeboat.*
39 *Lifeboat rocket blast shield.*
40 *Unpressurised cargo bay.*
41 *Port side cargo bay door.*
42 *Port passenger deck corridor.*
43 *Standard class passenger cabins.*
44 *Rear port manoeuvring/ docking rocket.*
45 *Hull plating impulse wave receptor panels.*
46 *Impulse wave power conduits linking hull plating receptor panels to accumulator cylinders via particle condensers.*
47 *Impulse power accumulator/ propellant cylinders.*
48 *Drive arm lift motor.*
49 *Main drive variable position support stanchion.*
50 *Impulse wave energy conduits.*
51 *Electro-sphere.*
52 *Ion particle accelerator.*
53 *Drive sphere rotation motor.*
54 *Steering/retro rockets. When decelerating, the drive sphere is raised above the level of the adjacent play dome and rotated to allow the rockets to face forward.*
55 *Rocket fuel tanks.*

DELAWARE LIFEBOAT

A *Avionics bay.*
B *Pilot and co-pilot position.*
C *Steering thrusters.*
D *Airlock.*
E *High powered long-range distress beacon.*
F *Seating for 42 passengers and crew.*
G *Personal hygiene station.*
H *Lifeboat bay access door.*
I *Air tanks and life-support system.*
J *Short-range port thrusters.*
K *Rocket fuel tank.*

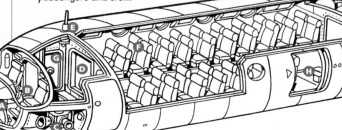

COMMERCIAL CRAFT

COMMERCIAL FREIGHT

In the early days, when space flight was an expensive and risky enterprise, most civilian companies sought government contracts, and very few attempted to become involved without such backing. However, one independent group, Cosmic Space Lines, took the gamble of going it alone, and not only survived to develop into a pioneering force in space – setting up the first interplanetary freight service – but also built their own research and development base on an asteroid, from which they eventually successfully achieved interstellar travel.

It was only when the twin developments of impulse drive and Hirshinium alloy came into widespread use in the early 1980s, making space travel much more affordable, that large numbers of civilian companies finally became involved in spaceflight, from one-ship independents up to large multinational corporations.

Lolo (London Luna Organisation) and SASS (South American Space Service) competed for the Lunar trade, during which SASS employed dirty tricks to sabotage Lolo's Diana 1 spacecraft after

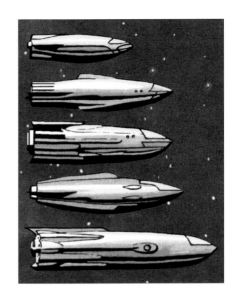

she had become the fastest ship on that route. The commercial and political fallout following the discovery of what SASS had done gave entrepreneur Simon Crowley the opportunity to take over SASS as the basis of his company Spacetours Incorporated. However, he was a greedy man and clashed with the ISF over the Venus–Earth food run in 1997.

The golden age of the independents was during the early food-run years, when every available ship was pressed into service to get vital foodstuffs to Earth. During this time they were able to make considerable profits, until Space Fleet was eventually able to expand its own heavy-duty fleet.

The Mekon's robot invasion of Earth in 2002 ended the food run, and by the time the Treen occupation ended in 2011 Earth's decimated population no longer needed extra supplies from Venus in order to survive. Commercial interest then shifted to terraforming Mars and exploiting the asteroid belt.

In 2016 civilian entrepreneurs' horizons grew even wider when the Halley and Nimbus space drives came into general use. Three years earlier Cosmic had led the way to the stars with its Galactic Galleon, and a year later the Nimbus showed that it could reach the outer planets with ease. Now others could follow.

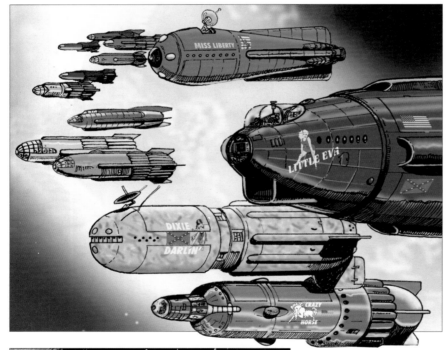

Top: Some Cosmic Space Lines craft.

Above: Civilian spacecraft on the Venus–Earth food run.

Left: Diana 1, fastest spaceship on the Moon run in the 1990s.

LUNABUSES

After the 2002–11 Robot War, Space Fleet took over responsibility for passenger fights to and from the Moon, and in 2012 it set up a special Lunabus fleet to service this route.

Designed to go direct from Earth to the Moon and back, rather than stopping at the over-used near Earth space stations (already busy with Mars and Venus traffic), the Lunabus is equally at home in Earth's atmosphere as it is in space. Not only can it carry 56 passengers and their luggage but its impressive cargo holds are also capable of supplying Moon Base with most of its off-world needs.

The craft are ramp-launched to save fuel, and behave like a graceful supersonic jet aircraft within Earth's atmosphere and a sleek spaceship when flying above it. Their sturdy double-hull construction – which even boasts a double nosecone as an extra protection against meteorites – has proved its worth, not a single life having been lost since these craft came into service. The fact that they have proved so effective in this role is the reason why, ten years on, they remain in active service.

ISF DATA FILE

When Moon Base was shut down in 2013 because all radio and radar wavelengths were being jammed by the Cosmobe fleet, a Lunabus made a textbook emergency landing in the adjoining Bonpland crater.

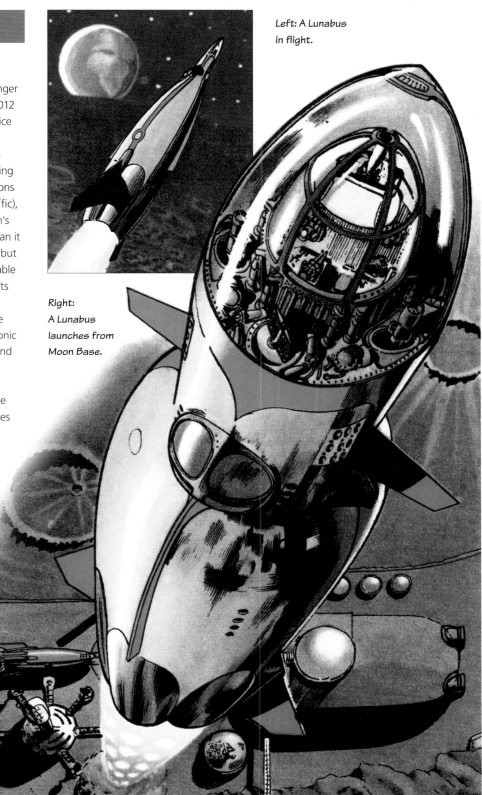

Left: A Lunabus in flight.

Right: A Lunabus launches from Moon Base.

LUNABUS

The Lunabus's direct service between the Earth and the Moon has a made it a popular and relatively affordable means of travel for both individuals as well as commercial enterprises using its cargo facilities. Its excellent safety record has ensured that the Lunabus service is both a profitable and reliable means of getting to and from Moonbase quickly and safely, and on a regular basis.

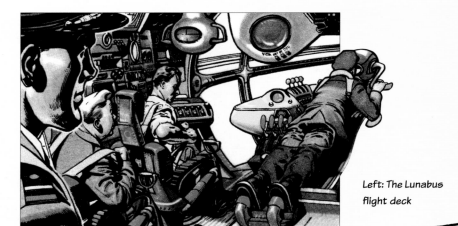

Left: The Lunabus flight deck

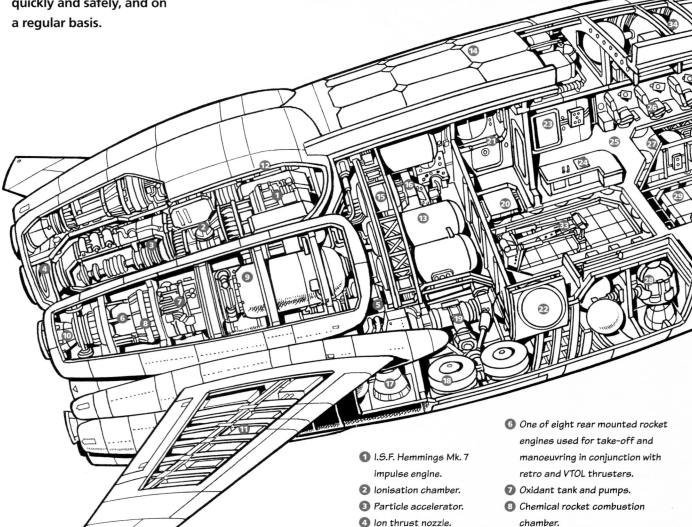

1 I.S.F. Hemmings Mk. 7 impulse engine.
2 Ionisation chamber.
3 Particle accelerator.
4 Ion thrust nozzle.
5 Impulse wave energy converter.
6 One of eight rear mounted rocket engines used for take-off and manoeuvring in conjunction with retro and VTOL thrusters.
7 Oxidant tank and pumps.
8 Chemical rocket combustion chamber.
9 Rocket fuel tank.
10 Chemical rocket thrust nozzle.

11 Impulse wave receptor panels built into rear mounted atmospheric re-entry stabilising tail fins.

12 Topside engine cowling containing two of eight take off and manoeuvring rockets.

13 Impulse wave accumulator/ propellant cylinders.

14 Impulse wave receptor panels.

15 Electrical power generating ring.

29 Underside cargo bay doors with integrated loading ramp.

30 VTOL rocket fuel tanks.

31 Port VTOL rocket tanks.

32 Forward starboard VTOL rocket thruster nozzle.

33 Cargo bay loading crane.

34 Compressed air tanks.

35 Water tank.

53 Flight controls.

54 Overhead flight controls and systems monitoring console.

55 Dual layered polarised canopy provides heat resistant protection during the atmospheric re-entry.

56 Radar, avionics and sensor arrays.

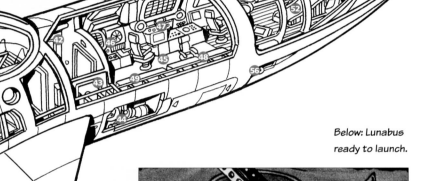

Below: Lunabus ready to launch.

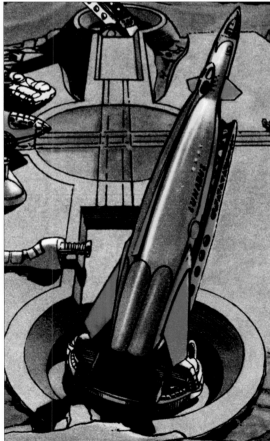

36 Life-support and atmosphere recycling and cleansing systems.

37 Crew-only access to cargo hold and supply bay.

38 Nosewheel.

39 First-class passenger saloon, seating up to 20 passengers.

40 Passenger saloon star field observation dome.

41 Atmospheric stabilising fins.

42 Flight deck equipment storage.

43 Crew toilets.

44 Forward retro rockets.

45 Flight deck.

46 Engineering station.

47 Navigation consoles.

48 Communications station.

49 Life-support control station.

50 Hand-operated sensor output monitor.

51 External environment, heat and radiation sensor.

52 Pilot's couch.

16 Engineering deck.

17 Rear starboard VTOL thrusters.

18 Starboard main wheel.

19 Telescopic main wheel strut.

20 Artificial gravity generator.

21 Port air lock.

22 Starboard passenger air lock outer hatch.

23 Passenger deck personal hygiene station.

24 Passenger deck bar.

25 Passenger desk, seating 36 passengers.

26 Passenger seating with TV consoles.

27 Passenger deck gravity plating.

28 Mid ships starboard VTOL thrusters.

SPACE BASES

SPACE FLEET HQ

By the time that Space Fleet came into being, the early rocket base developed alongside the former RAF Woodvale airfield at Formby Sands, Lancashire, was so well established that it was logical to make this Space Fleet HQ. Over the years since then the base has grown to include the RAF airfield as an expanded landing area, an offshore island for launching hazardous monatomic hydrogen (MH) spacecraft, a sea pier for recovery craft, and the Astral Cadet College at its south-east corner.

Above: Emergency at Space Fleet HQ in 2000.

ISF DATA FILE

When traffic became too great for one base to cope with it, a second spaceport was built near London along the same lines as the original base at Formby Sands. Most major planetary exploration flights took off from SFHQ, the exceptions being the Crypt and Cosmobe craft, which used the southern spaceport.

MOON BASE

Back in 1961, when space travel was expensive and Space Fleet had yet to be formed, the three principal spacefaring nations – Britain, France and the USA – pooled their resources to create a fledgling Moon Base, in order to explore and study Earth's satellite. Situated in the Parry crater, which borders the Fra Mauro and Bonpland craters, it took two years to build.

Once Space Fleet adopted impulse drive in 1976 an impulse relay station was built nearby, which quickly became very important, especially after 1980 when invention of the new lightweight alloy Hirshinium put development into overdrive. It was now feasible for civilian companies to venture into space, and there was huge public interest in the Moon. The implementation of landing fees was encouraged to cover the base's running costs, and a civilian terminal was built on the same site.

In 1998 a Treen electro-sender was added as a fast link between the base itself and the hydroponics farm/research facility on the northern side of the Parry crater and, beyond that on the other side of the crater wall, Balwin city, a new civilian development being constructed mostly underground (to shield it from solar flares and meteorites). To the south of Moon Base the electro-sender linked it with the impulse relay station. The aftermath of the 2002–11 Robot War changed everything. Space Fleet, now extremely security conscious, not only added a number of defence measures to the base but also restricted civilian access, and introduced a fleet of its own Lunabuses to cover the valuable tourist trade.

In 2014 a serious accident at a toxic waste disposal plant on Earth convinced the World Government that all such dangerous materials should henceforth be dealt with off-planet. The following year a special disposal plant was built alongside Moon Base to deal with such substances, using microorganisms found on Mercury that feed on radioactive material. The radioactivity of the waste is thereby neutralised, and the resulting residue transformed into an effective anti-radiation building material.

Above: Part of Moon Base in 2012.

ISF DATA FILE

In 2013 Moon Base had to close for over a week while the Cosmobe fleet passed through the Solar System. Their drive system disrupted all radio and radar wavelengths, which made landing spacecraft 'blind' too dangerous to risk.

MARSPORT

Marsport was developed from the original Chryse Planitia landing site in 1965. At first it consisted only of the two converted spacecraft that had been left behind during the first Mars expedition, but it continued to grow slowly and spasmodically right up until 1978, when an impulse station was constructed nearby. Thereafter expansion progressed much faster, especially once the new lightweight alloy Hirshinium appeared in 1980. Since it was now cheaper to get men and materials to Mars, construction commenced on a proper base next to the original site, and was completed in 1989. Commercial development of Mars followed – holiday centres sprouted up round the north pole, and geologists and mining companies began to probe and mine in many locations.

Marsport continued to grow until the Mekon's robot invasion in 2002. During the Treen occupation work and expansion were halted, but as Earth recovered from 2012 onwards so did Mars. By the time projects to terraform the planet got started in 2016 an entire linked dome city had grown up near the original Chryse Planitia base.

Above: View from Marsport control tower in 1988.

ISF DATA FILE

In 1987 the two original spacecraft around which Marsport had evolved were turned into a museum site.

McHOO ASTEROID BASE

The original McHoo asteroid base was built in secret in 1978 as a Cosmic Line research centre, by discredited Scottish scientist Halley McHoo. Most of it was destroyed in 1983 when the Galactic Pioneer took off, triggering a massive explosion. Halley McHoo was killed, but his son Galileo rebuilt and improved the base to continue his father's work. It now has two eight-berth mooring towers for deep-space craft, missile defences, and facilities for a huge interstellar spaceship on its far side. The hollowed out asteroid grows its own food and even keeps a herd of cows for fresh milk.

ISF DATA FILE

In 2013, just as the McHoo base was discovered by Space Fleet, the Galactic Galleon took off on what was to become the first ever round trip of a manned, Earth-built spacecraft to another star system.

Above: The McHoo asteroid.

Right: Part of the early McHoo base, just before the launch of the Galactic Pioneer.

1. Base security building.
2. Space Fleet officer's quarters.
3. Central Zone North Entrance 2 security barriers.
4. Domestic stores.
5. Base hospital and health centre.
6. Rocket service track.
7. Central Zone North Entrance 2 security barriers.
8. Ferry terminal.
9. Station engineer's office.
10. Launch ramp monitoring station.
11. Launch ramp blast duct, leading to reinforced filtered pipeline in the estuary.
12. Launch ramp.
13. Launch gantry.
14. Launch ramp in lowered position.
15. Space Fleet vehicle workshops.
16. Hangars No. 3 and 4.
17. Ferry hangar No. 5.
18. Freight loading cranes.
19. Freight offices.
20. Central Zone West Entrance.
21. Space Fleet HQ service road.
22. Formby beach; remodelled to accommodate service road, with no public access.
23. Landing stage.
24. Water filtration and purification plant.
25. Headquarters administration building.
26. Flight control building.
27. Workshops and stores located beneath rocket service track.
28. Rescue rocket launch silo.
29. Central Zone South Entrance 1 security barrier.
30. Meteorological station.
31. Research laboratories.

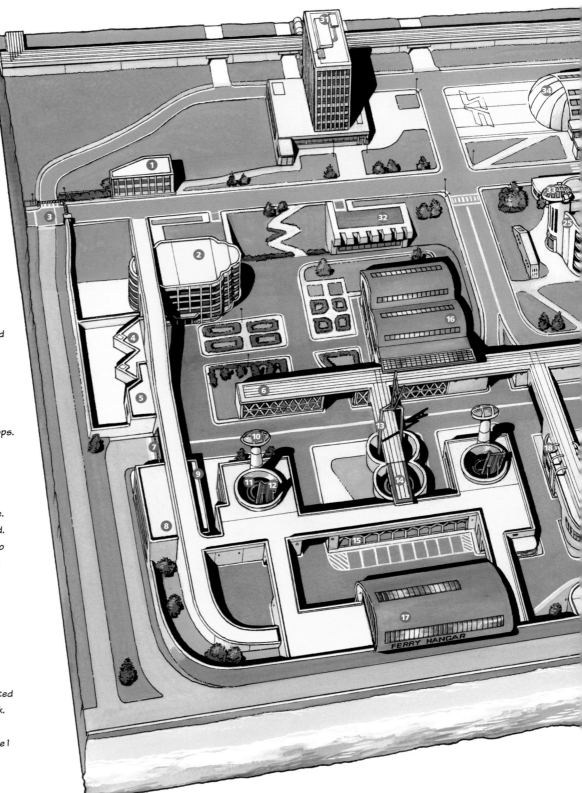

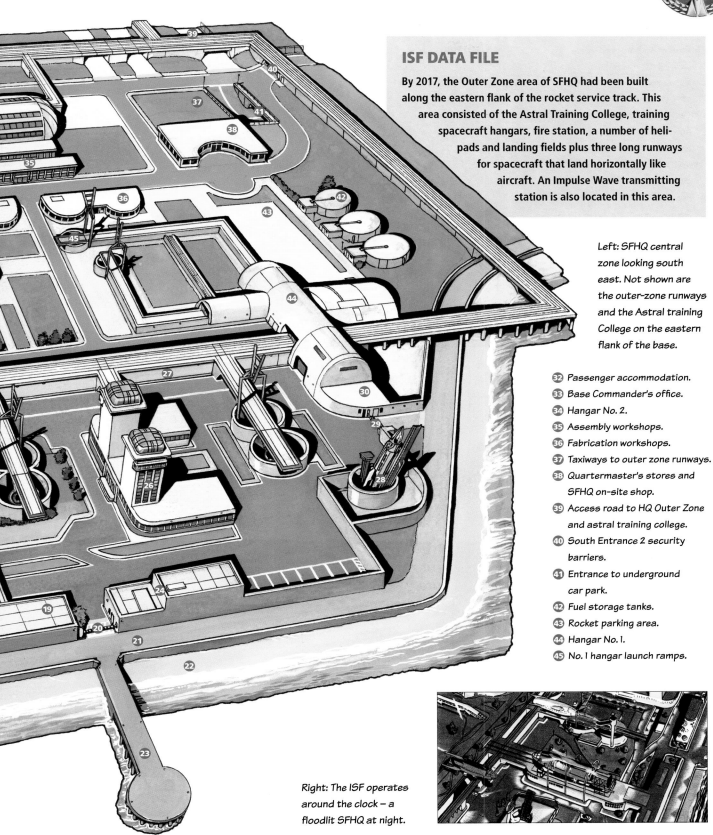

ISF DATA FILE

By 2017, the Outer Zone area of SFHQ had been built along the eastern flank of the rocket service track. This area consisted of the Astral Training College, training spacecraft hangars, fire station, a number of heli-pads and landing fields plus three long runways for spacecraft that land horizontally like aircraft. An Impulse Wave transmitting station is also located in this area.

Left: SFHQ central zone looking south east. Not shown are the outer-zone runways and the Astral training College on the eastern flank of the base.

32 *Passenger accommodation.*
33 *Base Commander's office.*
34 *Hangar No. 2.*
35 *Assembly workshops.*
36 *Fabrication workshops.*
37 *Taxiways to outer zone runways.*
38 *Quartermaster's stores and SFHQ on-site shop.*
39 *Access road to HQ Outer Zone and astral training college.*
40 *South Entrance 2 security barriers.*
41 *Entrance to underground car park.*
42 *Fuel storage tanks.*
43 *Rocket parking area.*
44 *Hangar No. 1.*
45 *No. 1 hangar launch ramps.*

Right: The ISF operates around the clock – a floodlit SFHQ at night.

MOONBASE

1. Hangar No. 1, specialising in spacecraft maintenance and repair.
2. Moonbase laboratories and workshop blocks covered by a radiation and meteorite-proof concrete shield.
3. Moonbase observation dome.
4. Airtight corridors linking the sub-surface laboratory blocks ensures personnel can access all parts of the Moonbase without the use of spacesuits.
5. Medical centre.
6. Moonbase hotel caters to the needs of space tourists providing additional income to help off-set the enormous running costs of Moonbase.
7. Heat, light and radiation-shielded tourist observation dome provides spectacular views of the lunar terrain.
8. Surface hangar atmosphere pumping and regulation systems.
9. Air-tight access corridors.
10. Spacecraft access airlocks.
11. Lunar ferry launch gantry in retracted position, allowing spacecraft to taxi into the adjacent hangar or other locations.
12. Gantry turntable.
13. Radar scanning operations dome.
14. Radar scanners' dual roles ensure early detection of incoming hostile spacecraft or missiles and also provides space traffic activity data around Moonbase relayed to central control.
15. Rocket service track.

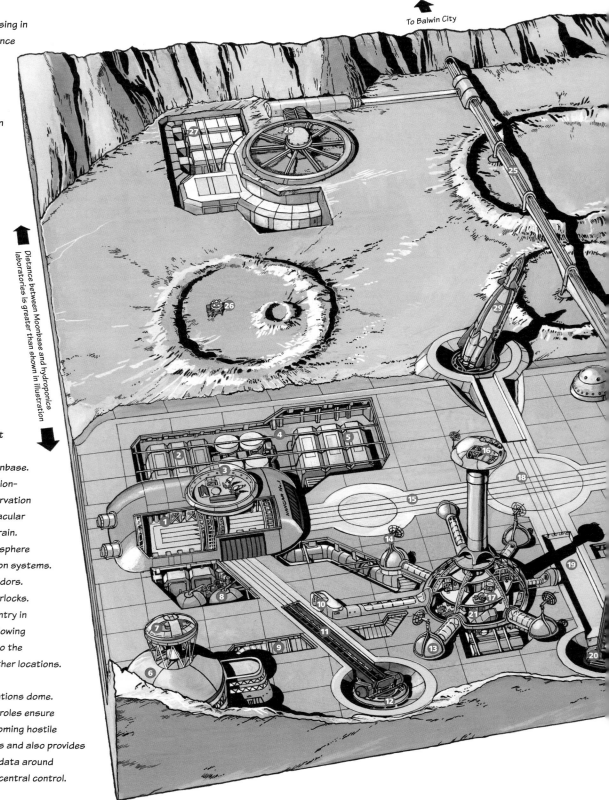

To Balwin City

Distance between Moonbase and hydroponics laboratories is greater than shown in illustration

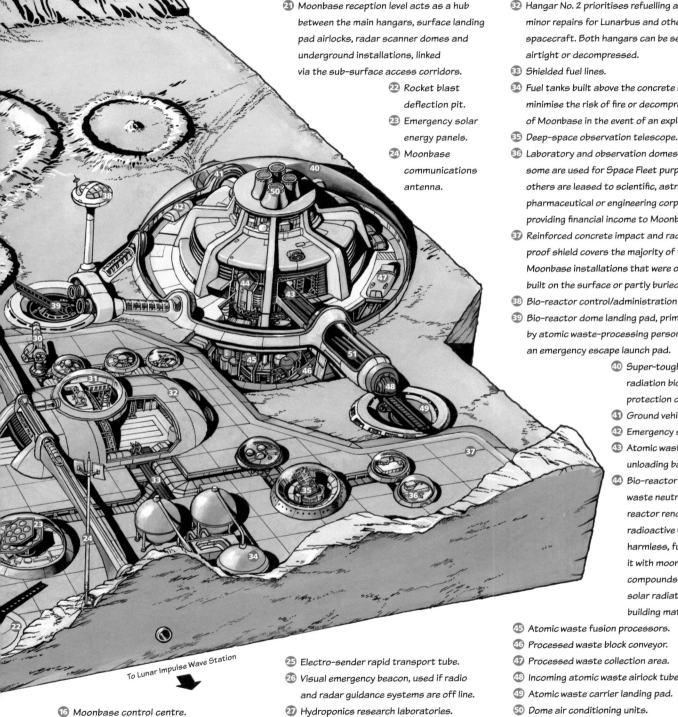

21 Moonbase reception level acts as a hub between the main hangars, surface landing pad airlocks, radar scanner domes and underground installations, linked via the sub-surface access corridors.

22 Rocket blast deflection pit.

23 Emergency solar energy panels.

24 Moonbase communications antenna.

32 Hangar No. 2 prioritises refuelling and minor repairs for Lunarbus and other small spacecraft. Both hangars can be sealed airtight or decompressed.

33 Shielded fuel lines.

34 Fuel tanks built above the concrete shield minimise the risk of fire or decompression of Moonbase in the event of an explosion.

35 Deep-space observation telescope.

36 Laboratory and observation domes. Whilst some are used for Space Fleet purposes, others are leased to scientific, astronomical, pharmaceutical or engineering corporations providing financial income to Moonbase.

37 Reinforced concrete impact and radiation-proof shield covers the majority of the original Moonbase installations that were originally built on the surface or partly buried.

38 Bio-reactor control/administration centre.

39 Bio-reactor dome landing pad, primarily used by atomic waste-processing personnel, or as an emergency escape launch pad.

40 Super-toughened anti-radiation bio-reactor protection dome.

41 Ground vehicle airlock.

42 Emergency shelter.

43 Atomic waste unloading bay.

44 Bio-reactor atomic waste neutralising reactor renders radioactive waste harmless, fusing it with moonrock compounds to create solar radiation-proof building materials.

45 Atomic waste fusion processors.

46 Processed waste block conveyor.

47 Processed waste collection area.

48 Incoming atomic waste airlock tube door.

49 Atomic waste carrier landing pad.

50 Dome air conditioning units.

51 Airlocked entry tube allows atomic-waste-carrying spacecraft to be brought into the bio-reactor protection dome so that the waste can be unloaded directly into the bio-reactor.

To Lunar Impulse Wave Station

16 Moonbase control centre.

17 Control dome administration level.

18 Spacecraft taxiing turntable.

19 Launch gantry stand-down well.

20 Lunar ferry launch gantry.

25 Electro-sender rapid transport tube.

26 Visual emergency beacon, used if radio and radar guidance systems are off line.

27 Hydroponics research laboratories.

28 Hydroponics farm.

29 Lunar ferry in launch position.

30 ISF long-range telescope.

31 Hangar control centre.

1. Mars colony landing strip control tower now disused but maintained in case of emergencies.
2. Disused maintenance hangars.
3. Mars colony landing strip. Now disused but maintained for emergency use.
4. Disused landing strip terminal.
5. Marsport personnel accommodation dome.
6. The Bradbury Dome; landing site of the first manned expedition to Mars. The dome encloses two of the three expedition spacecraft, exhibition areas and visitor centres.
7. Mars landing site commemorative sculpture.
8. Landing site museum.
9. Landing site workshops.
10. Exhibition hall.
11. Landing site visitors centre and cafe.
12. British Mars expedition spacecraft Winston Churchill rear section: the capsule was used to return the crew to Earth.
13. American Mars expedition spacecraft Franklin Roosevelt. Partly cannibalised on arrival to form temporary shelters. The ship has been rebuilt for visitors.
14. Dome lighting stanchion.
15. Bradbury dome life support and services area.
16. One of several airtight covered roads being built around Marsport to replace uncovered tracks and to supplement access corridors.
17. Space port fuel tanks.
18. Observation dome.

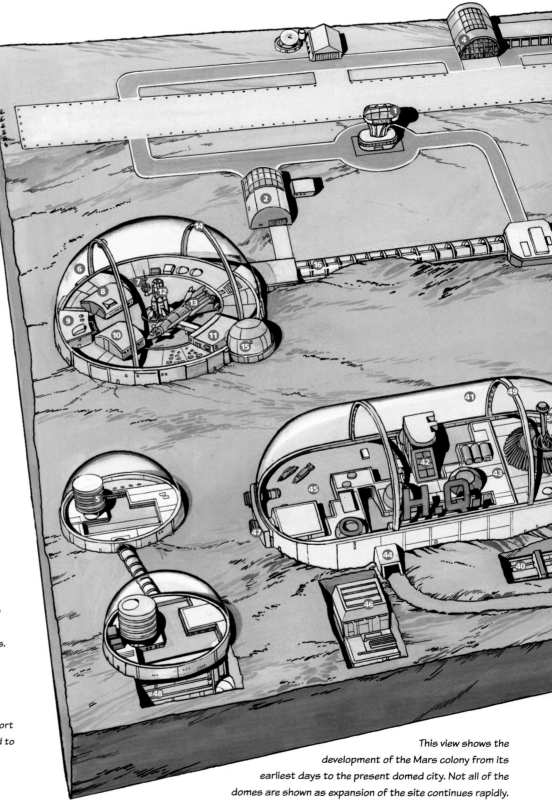

This view shows the development of the Mars colony from its earliest days to the present domed city. Not all of the domes are shown as expansion of the site continues rapidly.

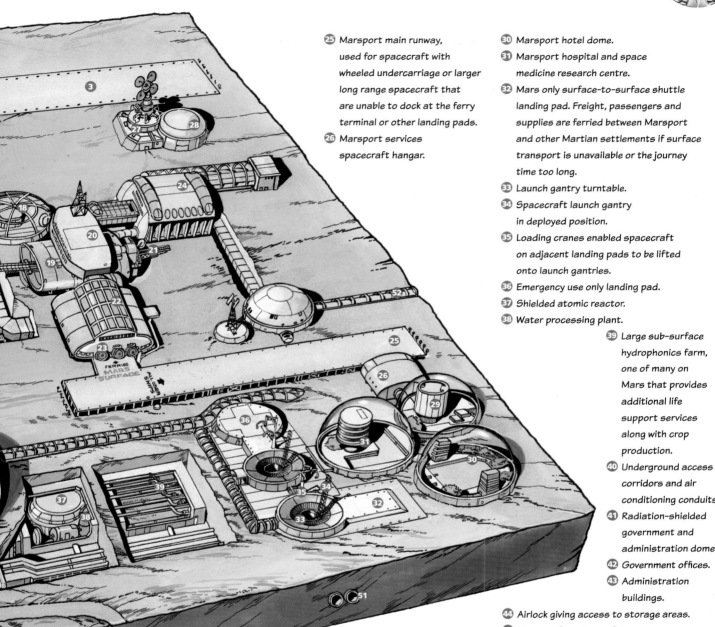

25 Marsport main runway, used for spacecraft with wheeled undercarriage or larger long range spacecraft that are unable to dock at the ferry terminal or other landing pads.

26 Marsport services spacecraft hangar.

30 Marsport hotel dome.

31 Marsport hospital and space medicine research centre.

32 Mars only surface-to-surface shuttle landing pad. Freight, passengers and supplies are ferried between Marsport and other Martian settlements if surface transport is unavailable or the journey time too long.

33 Launch gantry turntable.

34 Spacecraft launch gantry in deployed position.

35 Loading cranes enabled spacecraft on adjacent landing pads to be lifted onto launch gantries.

36 Emergency use only landing pad.

37 Shielded atomic reactor.

38 Water processing plant.

39 Large sub-surface hydrophonics farm, one of many on Mars that provides additional life support services along with crop production.

40 Underground access corridors and air conditioning conduits.

41 Radiation-shielded government and administration dome.

42 Government offices.

43 Administration buildings.

44 Airlock giving access to storage areas.

45 Mars surface-to-surface spacecraft servicing area.

46 Sub-surface emergency administration block.

47 Spacecraft airlocks.

48 Laboratories and workshops.

49 Dome support stanchion and lighting bar.

50 Deep-space scanner.

51 Launch gantry blast ducts.

52 Airtight surface-built road leading to further city domes and the Mars governor's residence.

19 Marsport departures.

20 Marsport space traffic control centre.

21 Ferry launch ramp.

22 Marsport ferry terminal hangar.

23 Mars ferry hangar airlocks.

24 Marsport passenger and freight terminal.

27 Impulse wave transmitting station, one of six on the Martian surface providing maximum deep space impulse beam coverage, allowing for Mars' rotation and solar orbit.

28 Dedicated impulse beam atomic reactor provides power for transmitting station.

29 Waste disposal, recycling and processing facility.

1. One of eight warning lights arranged around the surface of the asteroid. Each contains sensors and navigation beacons to ensure that McHoo's fleet operates safely around the base, particularly when docking.
2. Deep space scanner array.
3. McHoo spacecraft maintenance and storage hangar.
4. Maintenance vehicles and machinery used in the base's construction are stored in the hangar.
5. Spacecraft hangar outer airlocked doors (one shown in partly opened position).
6. Hydrophonics and food production farms.
7. Hydrophonics and life-support management systems.
8. Interceptor missile launch tubes.
9. Interceptor missile in deployed position.
10. Water recycling and purification plant.
11. Waste management and recycling systems.
12. Livestock management zones.
13. Docking bay for two spacecraft provides easy access to nearby laboratory dome.
14. Deep-space communications and radar scanner.
15. Laboratory dome.
16. Observation tower.
17. Base personnel transit system.
18. Separate and airlocked access passage.
19. Nuclear power reactor.
20. Beam guidance sensor array domes ensure safe docking of spacecraft at adjacent docking towers.
21. Control and base operations dome.
22. Guest accommodation.
23. Base transit system and its parallel access corridor each have gravity plating strips running their entire length, providing artificial gravity to personnel at every direction.

24. Telescope.
25. Personnel relaxation dome.
26. Power distribution station.
27. Central lift to docking tower base.
29. McHoo's throne room, the nerve centre of the asteroid base.

28. McHoo's accommodation control and administration dome; design inspired by Scottish castle motifs.

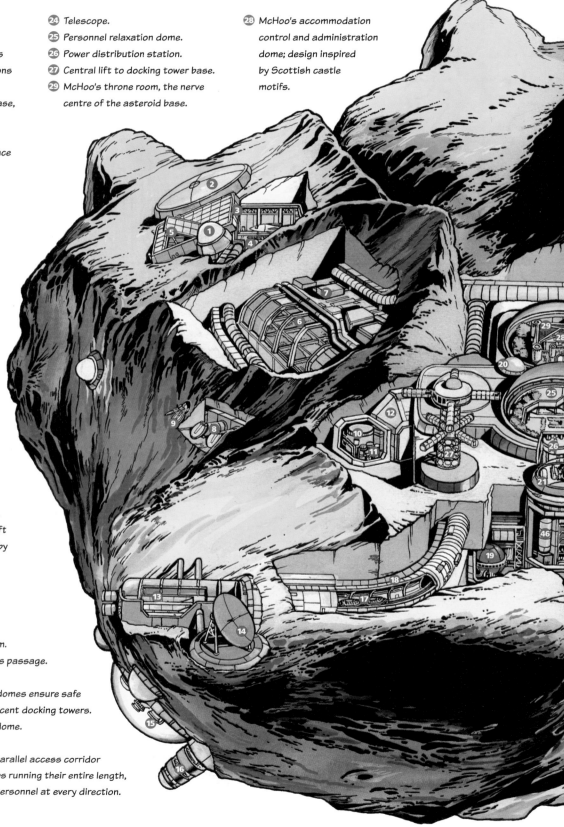

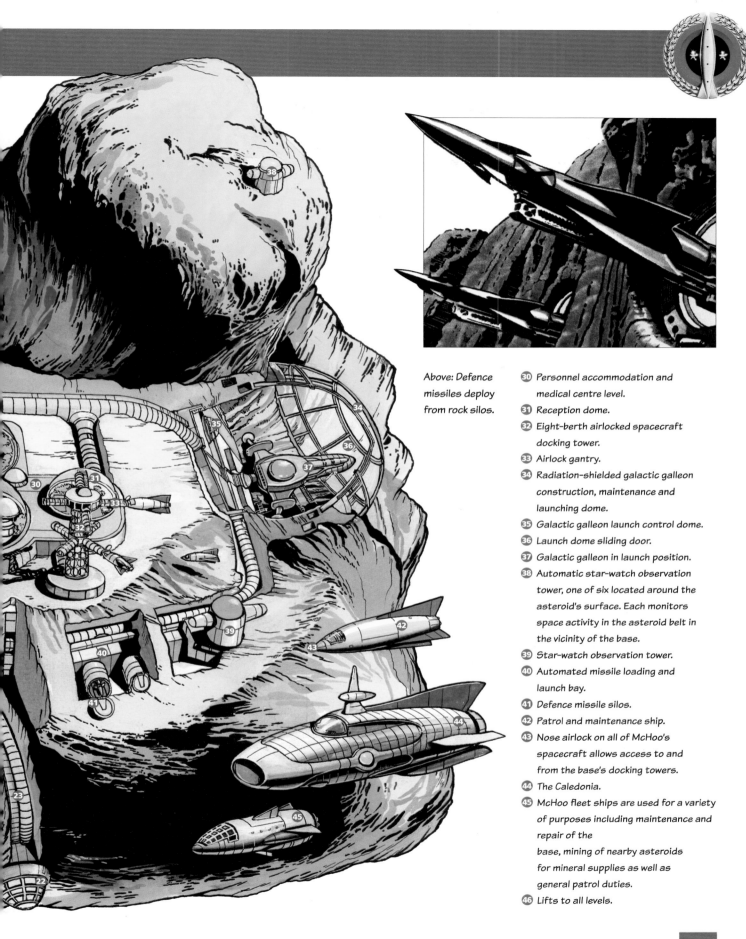

Above: Defence
missiles deploy
from rock silos.

30 Personnel accommodation and
medical centre level.

31 Reception dome.

32 Eight-berth airlocked spacecraft
docking tower.

33 Airlock gantry.

34 Radiation-shielded galactic galleon
construction, maintenance and
launching dome.

35 Galactic galleon launch control dome.

36 Launch dome sliding door.

37 Galactic galleon in launch position.

38 Automatic star-watch observation
tower, one of six located around the
asteroid's surface. Each monitors
space activity in the asteroid belt in
the vicinity of the base.

39 Star-watch observation tower.

40 Automated missile loading and
launch bay.

41 Defence missile silos.

42 Patrol and maintenance ship.

43 Nose airlock on all of McHoo's
spacecraft allows access to and
from the base's docking towers.

44 The Caledonia.

45 McHoo fleet ships are used for a variety
of purposes including maintenance and
repair of the
base, mining of nearby asteroids
for mineral supplies as well as
general patrol duties.

46 Lifts to all levels.

SPACE STATIONS

THE SPACE WHEEL

The Space Wheel, Earth's first proper satellite space station, remained in service into the 1990s, but was by then very outdated technologically. Its duties then passed to better, more advanced satellites, and it was just an abandoned shell when the Red Moon swept it away in October 1999.

Left: The Space Wheel space station.

THE 'J' SERIES SPACE STATIONS

The first 'J' series station was SFJ1, better known as 'Major One', one of three such stations completed in 1998. They were designed as termini both for short-haul ferry craft and the new deep-space ships which were then coming into service. Unlike the earlier Space Wheel, which had to spin to create its own artificial gravity, the SFJ stations had gravity motors under their decks, and ferries could actually dock inside the hull via twin airlock tubes. At the time of commission their groundbreaking facilities included a large domed restaurant with magnificent views of space and the deep-space craft docking platform that was above and slightly behind it.

Major One is stationed at the Lagrange point between Earth and the Moon, where the gravitational pulls of planet and moon are equalised, which is much further out into space than most

Above right: The upgraded SFJ3 in Earth orbit.

Above: Major One warns off spacecraft during the 'Communi-cations black-out' of 2012.

ISF DATA FILE

Major One had to close down for over a week during the communications 'black-out' in 2012, when the Cosmobe fleet was in the solar system.

other Earth-orbiting satellites. Its position saved it from being swept away in the wake of the Red Moon in 1999. As the spaceways became busier and more space stations sprang up nearer Earth, Major One became an emergency staging post for the Moon and deep-space expeditions.

SFJ2 was situated in close Mars orbit, with its departure ferry ramps pointing towards the Red Planet. Consequently in 1999 it became a victim of the Red Moon's fearsome magnetic pull as it circled Mars. Only the loading platform and evacuees were saved, the rest being dragged to destruction.

SFJ3 was stationed in close Venus orbit to assist with food transfers between Venus and Earth. Three years later the Treens built M.E.K.1 to help with the workload. After the 2002–11 occupation Earth's reduced population no longer needed so much from Venus, and M.E.K.1 was able to cope with the traffic unaided, so SFJ3 was towed to close Earth orbit and renamed Space Station Ten, a number of space stations having already been established near Earth by that time. In 2014 it was upgraded to deal with the new space ferries about to be introduced, and its arrival and departure areas were both rebuilt.

M.E.K.1

The Treens had no use for a space station until after the 1996 Earth–Venus war. Then a temporary one was pieced together around old Telezero spacecraft, to help facilitate the transfer of foodstuffs to Earth. This was supplemented in 1998 by Space Fleet's SFJ3. However, even together these two stations were hard-pressed to cope with all Earth–Venus traffic, so in 2001 the Treens built M.E.K.1 to replace their earlier stopgap station. M.E.K.1's central docking tower could handle four large deep-space craft at once, and in addition had 24 standard spacecraft docking airlock tubes located around its base. This sped up operations vastly, all governed by a magnificent 360° clear-domed control and monitor room that topped the central mooring tower.

Left: The M.E.K.1 in orbit around Venus.

Below: The Treens' temporary space station constructed from old Telezero ships.

Right: The Venus M.E.K.1 space station.

MARS 2

In 2000 the only 'L' type space station to be built replaced the destroyed SFJ2. It had two deep-space craft mooring piers, along with bigger and better passenger and cargo facilities than were found on earlier stations. This was fortuitous, since both Mars and the asteroid belt were attracting much greater attention from the mining and exploration communities, and Mars 2 became a supply depot for both.

Above: Two rotor cruisers docked at Mars 2, with a ferry leaving for Mars.

ISF DATA FILE

The Red Moon attack showed how vulnerable the Mars area was when its impulse power station failed, so plans were put in place to set up a network of impulse stations on evenly spaced asteroids around the inner rim of the asteroid belt's orbit field. That way not only would the whole of Mars' orbit be enclosed in an impulse net, independent of Mars, but the inner ring of the asteroid field would be covered too. Surveys were carried out in 2000, and work started the following year, but only two relay stations were operational when the robot takeover occurred in 2002. Construction resumed after the occupation ended, and the network was completed in 2012.

1. Ferry entry tube airlock.
2. Entry tube outer airlock door.
3. Ferry undercarriage guidance rail.
4. Ferry entry tube inner airlock door.
5. Maintenance airlock.
6. Ferry turntable: if required spacecraft can be turned through 180 degrees.
7. Turntable extension tracks enable ferries to enter the maintenance and refuelling deck via airlocked inner door.
8. Ferry passenger disembarkation area.
9. Airlocked inner door provides access for ferries to enter maintenance and refuelling deck.
10. Ferry maintenance and refuelling deck.
11. Astrodome passenger restaurant.
12. Restaurant food preparation area.
13. Restaurant bar.
14. Polarised radiation shielded restaurant dome.
15. Space station command centre.

16. Space station control room.
17. ISF video phone.
18. Engineer's office.
19. Station administration office.
20. Deep-space ISF communications antenna.
21. Impulse wave booster/relay station antenna.
22. Impulse wave mast warning lights.
23. Impulse wave relay booster power conduits.
24. Space observation and radar room.
25. Nuclear reactor provides power for the station's heat, light, gravity, life-support systems and the impulse wave booster station.
26. Compressed air tanks.
27. Air-recycling and life-support systems.

28. Artificial gravity generating ring: if required, gravity can be varied or switched off using deck plating gravity motor control systems.
29. Observation sphere.
30. Central access corridor.
31. Stairs leading to passenger baggage and cargo handling deck.

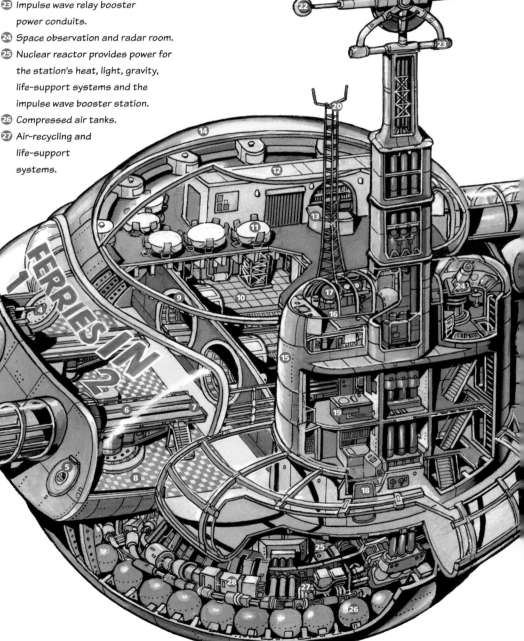

32 Mid-station ferry launching ramps.

33 Mid-station ferry launch tube incorporates a lift and turntable system allowing ferries to launch from the station's flank. Once in raised position, the tube is decompressed allowing the door to be opened and the ferry launched.

34 Mid-station ferry launch tube decompression pump and outer door mechanism.

35 Artificial gravity plating built into each deck.

36 Crew quarters.

37 Station water, waste and recycling systems.

38 Ferry air re-supply tanks.

39 Ferry re-fuelling tanks.

40 Station supplies and storage compartments.

41 Station positioning retro-rockets.

42 Passenger transfer lounge.

43 Food and drink dispenser.

44 Public video phone.

45 Passenger information desk.

46 Observation gallery.

47 Toilets and washroom.

48 Emergency escape airlock.

49 Loading platform lift.

50 Loading platform support tower incorporating lift travel guide tracks.

51 Loading platform support and stabilisation spars.

52 Retracted telescopic airlock.

53 Loading platform lift shaft is used by passengers as well as transferring luggage and cargo to and from the station to space trains and deep-space craft.

54 Loading platform control booth.

55 Loading platform gravity control console.

56 Airlocked telescopic docking tube door.

57 Inner docking tube airlock door.

58 Public access video and radio communications antennae.

M.E.K.I SPACE STATION

1 Radiation shield astrodome housing M.E.K.I control and observation room. Polarised quartz glass automatically reduces solar glare.

2 Control room central light.

3 Control room access stairs.

4 Long-range astrascope display screens.

5 Conference room.

6 Airlocked upper-level spaceship docking arm.

7 Docking-arm guidance beacon and searchlight.

8 Docking-arm manual/emergency control operating pod.

9 Hauser cable reel can be used manually to assist docking procedures.

10 Central mooring tower.

11 Administration offices.

12 Food and supplies storage deck.

13 Mooring tower utilities deck.

14 One of two airlocked lifts allowing access from the central mooring tower to the central operating zone. The telescopic lower section of each lift tube can be raised when the turntable is in use.

15 Turntable operating motor.

16 Central operating zone turntable. Spacecraft entering the central zone from one airlock can be transferred to another airlock in busy periods. Alternatively, craft can be turned 360 degrees to leave the station via the airlock it arrived through.

17 Cargo deck access doors.

18 Spacecraft servicing guidance tracks.

19 Localised variable gravity generator permits each airlock to vary its gravity intensity dependent on the type of spacecraft or its occupants using the station. Gravity control also aids spacecraft repairs and servicing.

20 Spacecraft repair and servicing cranes located within the airlock ring. Repairs to damaged craft can be carried out in weightless conditions (and in vacuum if required) before being taken into the space station's central operating zone. Cranes are used in conjunction with localised field generators and atmosphere control pumps/ blenders to optimise conditions for spacecraft personnel from different planets.

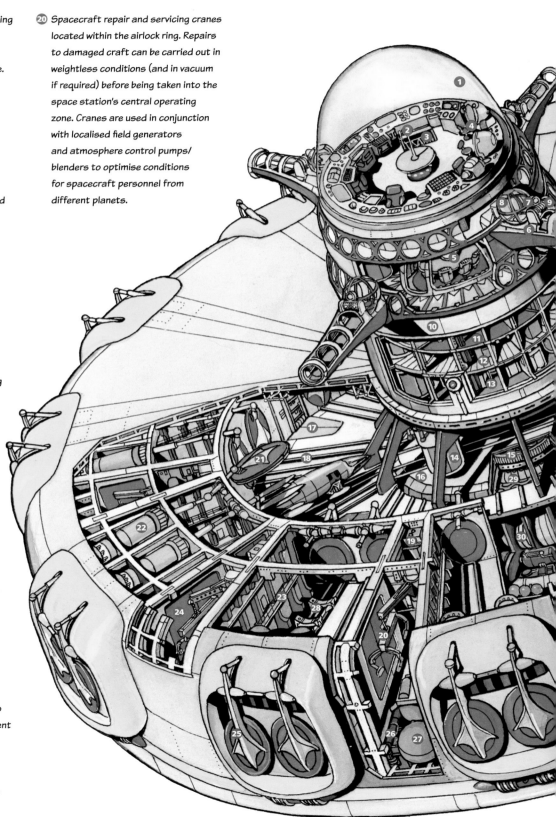

21 Inner airlock door.

22 Inner and outer airlock door control motors.

23 Airlock inspection gallery.

24 Servicing crane airlocked door.

25 Twin outer airlocked doors.

26 Airlock atmosphere pump.

27 Airlock atmosphere blender tanks.

32 Treen accommodation.

33 Treen nutrient baths.

34 Accommodation deck corridor encircles station.

35 Electricity generators powered by adjacent nuclear reactor.

36 Air tanks and life-support systems, linked to hydrophonics deck below (not shown).

37 Spacecraft refuelling tank. 24 individual tanks contain a variety of blended fuels for spacecraft of various models and origins. These tanks are prioritised for spacecraft that do not – or cannot – dock within the station itself.

38 Automatic fuel pump injector nozzle.

39 Scanner pylon.

40 Deep-space scanner dish revolves around the base of the station.

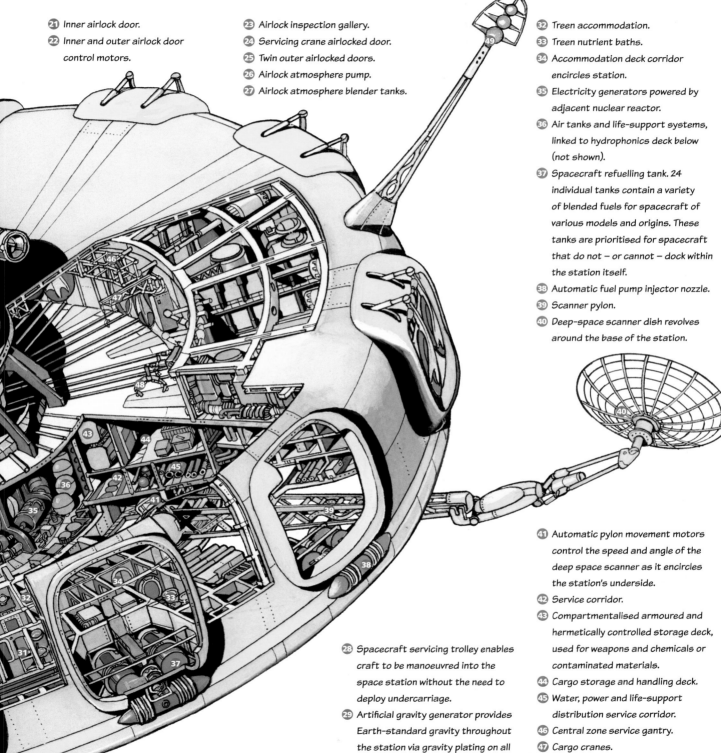

28 Spacecraft servicing trolley enables craft to be manoeuvred into the space station without the need to deploy undercarriage.

29 Artificial gravity generator provides Earth-standard gravity throughout the station via gravity plating on all floors, except in spacecraft airlocks.

30 Shielded nuclear reactor.

31 Crew cabins.

41 Automatic pylon movement motors control the speed and angle of the deep space scanner as it encircles the station's underside.

42 Service corridor.

43 Compartmentalised armoured and hermetically controlled storage deck, used for weapons and chemicals or contaminated materials.

44 Cargo storage and handling deck.

45 Water, power and life-support distribution service corridor.

46 Central zone service gantry.

47 Cargo cranes.

48 Cargo handling grabs.

49 Communication antenna and meteor detection system.

LATER SPACE STATIONS

'K' AND 'M' SERIES SPACE STATIONS

A final 'J' type space station – named ZQY, after the space coordinates where it was located – was built in 1999. The same year saw the first 'K' series stations being deployed. Painted bright yellow to make them easier to see in space, these improved and redesigned stations had the deep-space craft docking platform relocated below the structure, with larger airlock docking tubes running right through the station. Gone were the side-launching ramps of earlier series, as spacecraft could now enter the station via one end of the airlock tubes and exit via the other.

While ZQY and, later in 2012, the renamed SFJ3 station were put into close Earth orbit, a string of the newer satellites was placed further out to cover all approaches to Earth, and were letter-named accordingly.

Next came the 'M' class, mega-sized space station that included many elements of the 'K' design but on a much larger scale, along with major service and repair facilities. Spacecraft too badly damaged to survive planet-fall could now be retrieved rather than having to be abandoned in space. SFM1, nicknamed 'Big Bertie', was completed in near Earth orbit in early 2002, just before the Treen robot invasion.

In 2014, SFJ3 – now called Space Station Ten – was upgraded to take the newer designs of space ferries, but ZQY was shut down and later broken up.

Right: ZQY, the last 'J' type space station to be built.

Right: The Pescod attack on IQX.

ISF DATA FILE

A 'K' class satellite station, XQY was captured and later blown up by the Mekon and his Treen raiders in 2000.

Another 'K' type station, IQX, was destroyed by the 'Crimson Death' weapon in 2013, when the Pescod fleet passed by on its way to Earth.

While awaiting decommissioning, satellite station ZQY was taken over by a renegade scientist named Strombold, but he was defeated by Corky, an Astral College cadet.

Far left: The supersized 'M' class space station.

Left: Inside the 'M' class space station.

MARS MAINTENANCE SATELLITE

Once the asteroid impulse net was completed in 2012 commercial interest in the inner asteroid belt expanded exponentially, so that the Mars 2 station was unable to cope with the steadily increasing rise in traffic to both Mars and the asteroid belt. Consequently the Mars Maintenance satellite was constructed and put into Mars orbit to supply exploration and mining craft working the belt, leaving Mars 2 to deal with traffic heading directly for the Red Planet. Mars Maintenance therefore didn't have any passenger facilities, viewing rooms or airlock tubes running through it. Instead it had eight docking pods attached to its sides with access to its onboard supply of stores, equipment and spares.

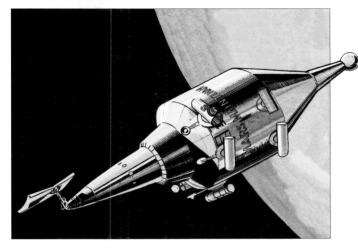

Above: The Mars Maintenance Satellite.

Left: Mars Two space station.

ISF DATA FILE

In 2014 Mars Maintenance was the scene of the dramatic capture of criminal Martin Brand and his accomplice in what would become known as 'the Solid Gold Asteroid caper'.

SPA-ONE

The appearance of the Cosmobe and Pescod in the inner Solar System in 2013 was the last straw as far as the World Government was concerned. Having been caught out a number of times by alien craft appearing suddenly near Earth, they decided to supplement the existing inner defence network with a powerful radio/radar dish to detect impending threats long before they drew too close. Therefore with the aid of Space Fleet they set about the construction of Spa-One, a major orbital space observatory, which was completed in 2014.

ISF DATA FILE

Shortly after going operational Spa-One was instrumental in locating the missing crew of Nimbus 1 when aliens stranded them in space near Jupiter. It also located and tracked the aliens' craft so that Nimbus 2 was able to intercept it.

Above right: Spa-One in Earth orbit.

Right: A high-speed ferry launches from Spa-One's space dock.

DEFENDING EARTH

When Space Fleet was formed in 1966 there was no expectation of encountering any sort of threat from outer space, so its Earth bases were defended only by UN fighter aircraft, ground-based missiles and conventional troops. When, 30 years later, the Treen menace appeared it was thought to have been stifled on Venus, so very little was done to update the defensive capabilities of Space Fleet's bases. Four years after this the inadequacy of its defences became only too obvious when unfriendly Kroopak ('Black Cat') craft attacked Earth.

As a result Elite Squadron was formed, consisting of six three-man, missile-armed 'star' spacecraft, operating in conjunction with an eight-man command ship, the Speedstar.

Left: Mobile missile launchers deployed at SFHQ.

Left: UN fighter aircraft chasing 'Black Cats'.

ISF DATA FILE

Elite Squadron were scrambled to contain the Mekon and his renegade Treens when he captured the XQY space station in 2001, but arrived too late. However, they pursued him to Venus, suppressed the Treens who tried to assist him and captured the Mekon after his spacecraft was destroyed by a mine.

In 2001 the first specially designed interceptor ships came into service – the SI Mark 2 (the Mark 1 never made it off the drawing board). These were fast three-man craft, armed with two swivel-mounted disintegrator cannon, one on each side. They were much faster than the Elite ships and became the backbone of Earth's inner defence system. Often operating in groups of three or six, they were launched from purpose-built three-ship ramps.

After the 2002–11 Robot War, Earth's defences were further strengthened when a network of remote-controlled missile satellites was set up around the planet as an outer perimeter shield. Unfortunately the shield failed to stop the Pescod fleet in 2013.

Above: Elite Squadron leaves Earth.

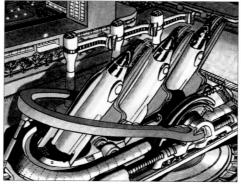

Right: An SI triple launch ramp.

Left: Missiles launch from a UN defence satellite.

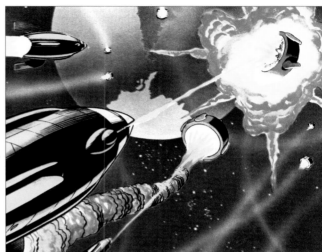

Above:
XC-9 Space
Cruisers.

Above right:
Mark 4s in
action against
the Triton fleet.

Left: Blue
Squadron
launching.

Below: The
new Galaxy-
Class battle
cruiser.

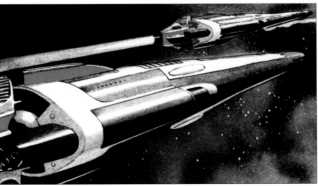

Space fleet now had its inner defence system. An experimental Z-19 craft (see page 106) recovered from the Sargasso Sea of Space showed that Proto drive did indeed live up to its promise, so more were commissioned for long-range patrol work. However, when shortly afterwards a new cruiser design was tested – the XC-9, which also used Proto drive – it was found to eclipse the Z-19 in several ways, and was rapidly adopted by the fleet as its main long-range patrol craft.

In 2017 'Blue Squadron' – an experimental wing of one-man rocket interceptor ships entered service, but were withdrawn in 2018

In 2018 the much superior SI Mark 4 entered service, its offensive capability enhanced by four concealed forward-facing missile launchers. In addition the cannon blisters could now be operated via a cockpit targeting panel. The cockpit itself was redesigned to function as a survival capsule in the event of an emergency or if the ship suffered severe battle damage. In such situations it can separate from the rest of the ship, automatically activating a distress beacon in the process. Alternatively, if it is in a planetary atmosphere parachutes will activate and bring it safety down to the ground.

The latest type of outer system patrol craft – the Galaxy-Class battlecruiser – has only just come into service, to safeguard approaches to the inner planets. Many of the details remain classified, but it can be confirmed that they have the latest photon drive technology, and that long-range nuclear torpedo weapons are included in their arsenal.

Mk4 SPACE INTERCEPTOR

Having already proved itself in two major actions, the Mark 4 Space Interceptor has emerged as the best of its kind. As well as incorporating all of the most successful features of the earlier versions, it also has a versatile weapons system, and its cockpit can double as a survival capsule if battle damage prevents it from being able to land. Nicknamed 'The Yellow Peril', it is very popular among the crews that fly it.

Above: A Mark 4 with its retro-rockets firing.

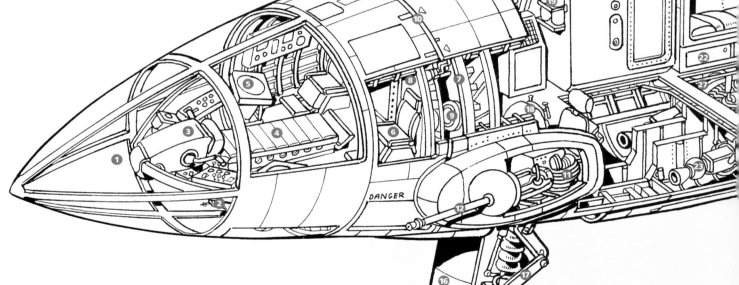

1. Armoured heat and radiation-proof, polarised nose cone viewport.
2. Scanner and avionics bay in lower cockpit.
3. Primary spacecraft flight controls. Port and starboard cannons can be operated from this console if required.
4. Pilot's couch.
5. Navigation display screen.
6. Crew acceleration couches.
7. Main body bulkhead.
8. Nose cone sealed bulkhead.
9. Nose cone attachment joints.
10. Nose cone emergency detachment explosive bolt cover.
11. Disintegrator cannons charging unit.
12. Forward port disintegrator cannon in deployed position.
13. Forward topside retro rocket.
14. Armoured rocket fuel tank housing.
15. Forward nose wheel.
16. Nosewheel airtight door.
17. Torque scissor links.
18. Port gunnery officer's station.
19. Fire extinguisher.
20. Water tank.
21. Personal hygiene station.
22. Medical supplies storage.
23. Gimballed underside pitch and yaw/retro rockets.
24. Medical bay is equipped for emergency treatment and minor operations.
25. Medical light and scanner.
26. Scanner display screen.
27. Artificial gravity generator.

28 Gravity deck plating.
29 Port retro rocket.
30 Hinged panel opens when retro rocket is in use.
31 Air-conditioning control system.

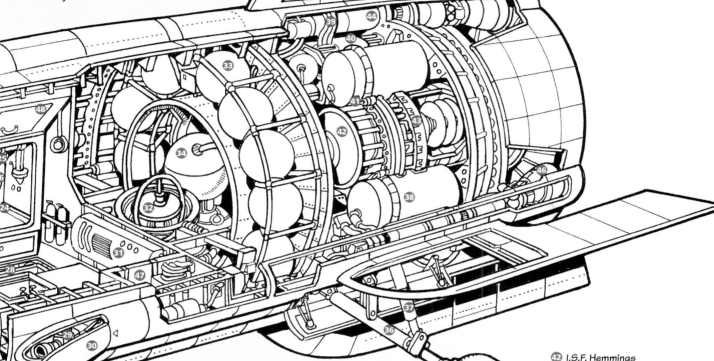

32 Attitude adjustment gryroscope.
33 Compressed air tanks.
34 Air scrubber and life-support system.
35 Rear port landing wheel.

36 Port wheel landing strut.
37 Landing strut drag brace.
38 Impulse wave accumulator tanks.
39 Impulse wave power conduits.
40 Impulse wave power converter.
41 Charged impulse wave distribution valves.

42 I.S.F. Hemmings Mk 9 impulse engine.
43 Ion particle accelerator.
44 Topside thruster rocket.
45 Impulse wave receptor panels built into all four tail fins.
46 Port thrusters rocket nozzle.
47 Spacesuit storage locker.
48 Storage lockers.

NOSE CONE ESCAPE PROCEDURE

In an emergency, the nosecone can detach using explosive bolts, either to await rescue or descend to Earth by parachute. Life-support systems are, however, limited so rescue or landfall is recommended within 12 hours.

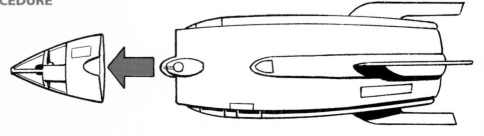

Mk4 SPACE INTERCEPTOR

61

XC-9 SPACE CRUISER

Built around a fast efficient Proto drive, these long-range ships have tough double Hirshinium/tungsten plate armour and are armed with four swivelling disintegrator cannon blisters evenly spaced around the hull, along with forward-facing missile launchers. Designed for long deep-space patrols, they normally carry a three-shift crew totalling 27 men, with bunking for nine at a time. To assist its space policing duties the XC-9 has an in-space concertina docking tube that enables it to link with and board other ships. In addition it carries two two-man 'space chariot' inspection craft.

Above: Side-on view.

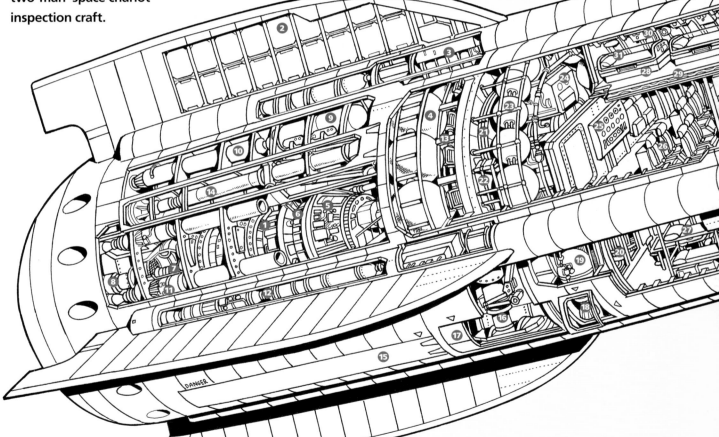

1. I.S.F Hemmings Mk. 18 Impulse Wave 'Proto drive' ion engine.
2. Impulse wave receptor panels.
3. Impulse wave power conduits.
4. Impulse wave accumulator/propellant tanks.
5. Ion drive electro-sphere.
6. Ionisation chamber.
7. Particle accelerator.
8. Ion thrust nozzle.
9. Thruster rocket fuel tank.

10 Thruster rocket oxidant tank and pumps.

11 Thruster rocket combustion chamber.

12 Rocket exhaust outlets.

13 Electrical power systems generating ring.

24 Air-recycling and life-support system.

25 Life-support controls.

26 Crew acceleration seats.

27 Crew quarters: 9 bunks provide sleeping accommodation for 27 crew members on a 3-shift duty rotation.

43 Middle deck gravity plating.

44 Personal hygiene station.

45 Underside disintegrator cannon.

46 Forward starboard vertical thruster.

47 Starboard rocket fuel and oxidant tanks.

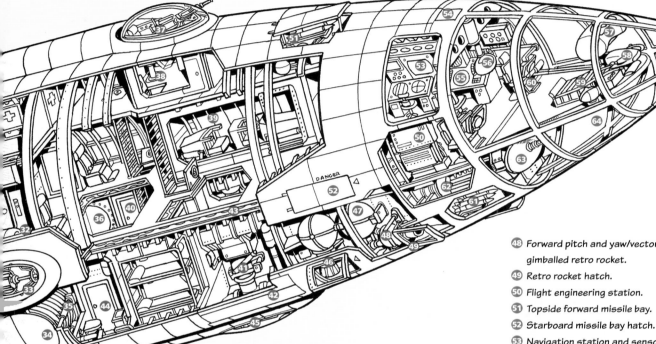

48 Forward pitch and yaw/vectoring gimballed retro rocket.

49 Retro rocket hatch.

50 Flight engineering station.

51 Topside forward missile bay.

52 Starboard missile bay hatch.

53 Navigation station and sensor output monitors.

54 Long-range camera hatch.

55 Communications station.

56 Hand operated forward scanner controls displays output from radar scanner, sensors and long-range camera above the flight deck.

57 Overhead flight status display console.

58 Flight controls.

59 Pilot's couch.

60 Commanding officer's position.

61 Underside missile bay.

62 Avionics bay.

63 Radar scanners and sensor array.

64 Polarised armoured flight-deck canopy, bonded with ceramic additives is heat and radiation resistant.

14 One of four rear mounted landing legs enable the ship to land in an upright position if required.

15 Landing leg door.

16 Rear starboard retracted landing leg.

17 Landing leg door.

18 Rear starboard vertical thruster.

19 VTOL rocket fuel tank.

20 Artificial gravity generator controls deck plating and gravity motors on all levels.

21 Gyroscope controls ship's pitch and yaw operations in conjunction with retro rockets.

22 Engineering and life-support systems access corridor.

23 Compressed air tanks.

28 Medical bay.

29 Upper deck gravity plating.

30 Patient injury monitoring systems.

31 Sealed medical bay beds secure patients during ship's combat manoeuvres.

32 Starboard gunnery officer's station.

33 Starboard disintegrator cannon.

34 Starboard airlock outer hatch.

35 Telescopic airlock tube deployment/ retraction motors.

36 Port airlock inner door.

37 Topside disintegrator cannon.

38 Topside gunnery officer's station.

39 Port space chariot holding bay.

40 Port hygiene station.

41 Forward starboard retracted landing leg.

42 Landing leg door.

SPACE FLEET STANDARD SPACE SUIT

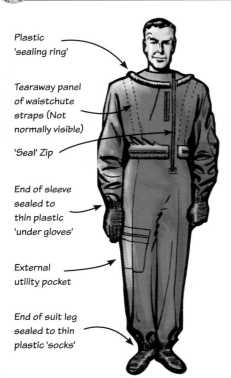

Plastic 'sealing ring'

Tearaway panel of waistchute straps (Not normally visible)

'Seal' Zip

End of sleeve sealed to thin plastic 'under gloves'

External utility pocket

End of suit leg sealed to thin plastic 'socks'

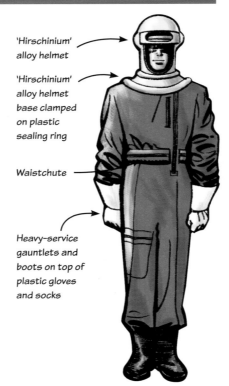

'Hirschinium' alloy helmet

'Hirschinium' alloy helmet base clamped on plastic sealing ring

Waistchute

Heavy-service gauntlets and boots on top of plastic gloves and socks

ISF DATA FILE

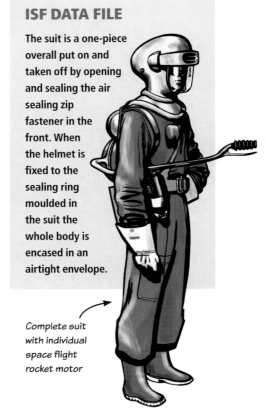

The suit is a one-piece overall put on and taken off by opening and sealing the air sealing zip fastener in the front. When the helmet is fixed to the sealing ring moulded in the suit the whole body is encased in an airtight envelope.

Complete suit with individual space flight rocket motor

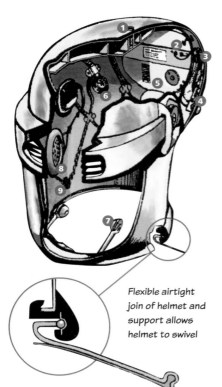

Flexible airtight join of helmet and support allows helmet to swivel

The helmet is the nerve centre of the suit. Constructed of Hirschinium – a strong, lightweight, non-conducting alloy, perfected by Erwin Hirshbaum in 1980. In the rear of the helmet, in the 'overhang' incorporated to carry them, electronic control devices ❷ keep the wearer comfortable inside the suit at all times. A thermostatic control ❸ maintains temperature through a network of heating wires ❾ and cooling channels embedded in the suit fabric. A diaphragm valve ❶ connects with an outlet valve in the air tanks built in the helmet crown. The tanks store a liquified combination of breathable gases at an incredibly low temperature, and are insulated from the rest of the suit by the non-conducting Hirschinium. This liquid breathing mixture is released into the suit through the outlet valve. This warms and expands it through the Thermovent device ❸, invented in 1985 by Lewell Hudson. The diaphragm regulates the gas to match the pressure of

any 'reasonable' atmosphere. In space the diaphragm is naturally at Minimum Pressure, which releases just enough for breathing.

Some helmets are equipped with auxiliary air-intakes in the rear overhang, which draw in atmospheric air through filter pads ❻. The filters work much like a poison gas respirator, passing through a breathable combination.

Spring-loaded throat microphones ❼ transmit speech on short-range radio ❺ to other suit wearers via built-in earphones ❽. Long-range radio transmission can be used by plugging a 'Walkie Talkie' set into the 2-pin plug ❹ on the helmet.

A storage pocket is normally used for carrying tools, but during exploration it carries a Hyposyringe kit and water-making outfit. In an emergency, it is possible to survive for an extended period within the suit by taking food in the form of injections administered by the Hyposyringe directly through the self-sealing fabric of the suit.

Above: Close-up of multi-layered cage crash helmet.

Lightweight Mars suit

The lightweight Mars suit with a goldfish-bowl helmet. Designed to overcome the thin Martian atmosphere, these suits can cope with temperatures as low as minus 100 degrees F.

The heavy-duty crash kit

For test flying new and potentially dangerous craft a special heavy-duty crash kit is worn with extra think armour plating covering the whole body plus a multi-layered cage crash helmet.

The battle suit

A specially armoured combat suit. The fabric has an extra super-tough layer to resist penetration. The suit comes with thick knee-length armoured boots and armoured waist, groin and upper thigh panels.

At times Space Fleet has used a number of other suits. In addition to those shown here, there was a synthetic rubber version of the Mark 4 suit (with the helmet sprayed with a film of transparent rubber) to protect personal from the Crimson Death weapon encountered when the Pescods landed on Earth in 2013.

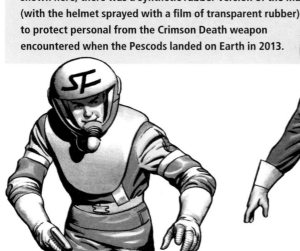

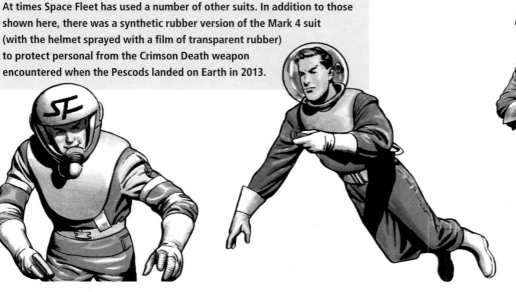

2014 type suit

A new variety of space suit was introduced in 2014, using synthetic fabrics and redesigned helmet. However, it was later found that the fabrics stiffened after repeated exposure to vacuum, and the suit was removed from service.

2015-2016 type suit

Following the failure of the 2014 suit, various designs were trialled between 2015 and 2017 combining both old and new technology. Shown here is the fish-bowl helmet type.

Later 2016 type suit

Another variation of space suit design trialled in 2016. Eventually it was decided to return to the successful designs of earlier years: the standard Mark 4 space suit, and the battle version of the same.

SPACE FLEET EQUIPMENT

SPACE CHARIOTS

Space chariots were first introduced to help the satellite service teams, by providing them with a better way to carry out inspections and repairs in space. The standard variety, shown here, can seat two, can anchor itself to a work area and can tow toolboxes or spare parts behind it. Bigger spacecraft often carry a space chariot for emergency inspections, or to transfer personnel to other craft in deep space.

There have been one or two variations on the chariot, including single-seater buggies. The Nimbus craft tried out the rounded versions (see page 116), and a longer 'bus' version capable of transporting five people is now beginning to appear.

Left: Searchlight in use.

Below: Chariots using magnetic nose clamps on a ship's side.

Right: View of steering handles and boot clamps.

1. Electro-magnetic suction clamp.
2. Forward search light and proximity sensor.
3. Proximity sensor computer, used in conjunction with electromagnetic clamp.
4. Proximity sensor and search light control wire conduits.
5. Forward starboard retro rocket.
6. Forward port multi-plane manoeuvring thruster nozzle. Four rockets control pitch and yaw/attitude course corrections along with forward and reverse thrust.

7. Suction clamp generator and back-up control systems generator.
8. Forward port retro rocket.
9. Rocket fuel lines leading to under seat fuel tanks.

10. Flight control systems computer.
11. Pilot's instrument console.
12. Steering controls.

13. Space suit boot clamps help ensure chariot operators remain in position.

14. Seating sidebars help ensure chariot operators don't drift away from the vehicle in the weightlessness of space.
15. Pilot's seat.
16. Electric power generator and rechargeable power cell.
17. Secondary operator's space suit boot clamp.
18. Rocket fuel tanks.

19. Secondary chariot operator's seat; often occupied by a specialist engineer or medical operative.
20. Tool compartment access panel.
21. Hand-held tool compartment.
22. Grappling hook, can be operated by hand or from the console
23. Hauser cable.
24. Hauser cable deployment/ rewind motor.
25. Rear port multi-plane manoeuvring thruster nozzle.
26. Medical supplies and equipment storage.

ISF DATA FILE

The Marco Polo exploration ship carries ten space chariots, as she was designed to investigate the asteroid belt. This large complement of chariots proved invaluable when she was sent to map the Sargasso Sea of Space.

LAUNCH GANTRIES AND LOADING CRANES

1. Gravity-adjustable spacecraft launch ramp. Gravity plating built into the launch ramps generates limited gravity fields enabling the smooth launch of spacecraft leaving the gantry. This prevents damage to the underside of the vessel as it takes off.

2. Ramp gravity generators.

3. One of two gantry hydraulic support stanchions.

4. Launch tower assists spacecraft lift-off procedures using gravity stabilisation fields to ensure a smooth launch.

5. Spacecraft tailfin slot.

6. Gravity stabilisation field generator maintains spacecraft's position on the gantry prior to and during launch.

7. Tower protection blast shields.

8. Telescopic spacecraft embarkation corridor adjusts the entry point for boarding crew depending on the position of the airlock on spacecraft of different design.

9. Gravity lock built into embarkation corridor floor. Used in conjunction with the spacecraft's own gravity systems, gravity locks enable boarding crew to be re-orientated with the vessel's own artificially generated gravity direction prior to lift off.

10. Embarkation tower is height adjustable to allow crew to board spacecraft with differing airlock configurations.

11. Emergency steps.

12. Access lift to spacecraft embarkation corridor.

13. Rocket service track shown in lowered position prior to spacecraft launch.

14. Blast doors open during launch operations to allow rocket exhaust to enter blast duct below.

15. Blast duct grille.

16. Blast duct leading to undersea outlets within the Space fleet HQ-controlled coastal zone.

17. Launch gantry powerhouse gravity control consoles and monitoring systems.

18. Launch gantry gravity control power turbine.

19. Blast and soundproof launch gantry power house roof.

20. Rocket service track.

21. Guide track traction beam generators.

22. Guide track power conduits.

23. Traction beam receiver maintains vehicle's stability on the service track.

24. Rocket service track spaceship towing vehicle.

25. Driver's cabin.

26. Computer-controlled independently steered hover nacelles.

27. Rocket towing hook.

28. Gravity-locked guide tracks. Each individual track generates a gravity traction beam to receivers on the underside of each service vehicle, locking them on to the rocket service track during spacecraft towing and maintenance operations.

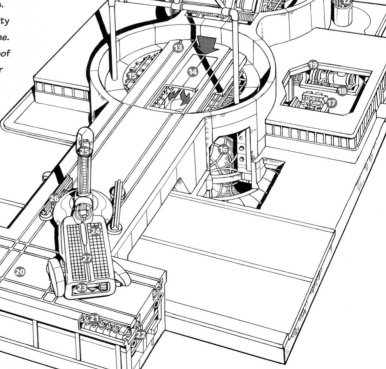

SPACE FLEET WEAPONS

Space Fleet was originally equipped with civilian weapons, as it was thought that the most they would be called upon to do was assist riot control police, or deal with occasional intrusions by protesters. Therefore paralysing guns were the issue of the day: a Hemming Mark II pistol for close action, and a Versun Tupper rifle, firing paragas pellets, for longer-range use. After the 1996 Venus war, a version of the Treen flame gun was added to the

heavy-duty space kits, but more as a maintenance tool than a weapon At the time of the Saturn mission in 2000 Space Fleet crewmen had autoray guns. However, their use was short-lived, as the Wyndham-Clark disintegron rifle became the standard combat weapon the following year. In 2014 the Aldiss electro-stunner came off the experimental list to serve as an alternative non-lethal handgun alongside the paralysing pistol.

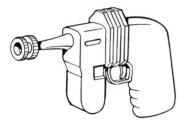

Standard paralysing pistol
The small clip-in pressurised paragas canister used in this gun is good for ten two-second bursts of gas.

Autoray rifle
This automatic ray gun fires bursts of brilliant narrow-beam rays. The intention was to temporarily blind and disorientate opponents, but it was found that the light rays could sometimes be reflected back at the user.

Flame blaster
This is really an engineer's tool with three settings. On its lowest setting it can be used to free a frozen switch, but on its highest setting it will melt metal.

Versun Tupper paragas pellet rifle
A weapon that uses magazines of 20 paragas pellets that clip into the rear handgrip. It takes about five seconds to swap magazines. A very useful weapon for riot control.

Electro-stunner gun
An impulse-wave powered gun that delivers a directed non-fatal electric shock. It can keep firing for as long as it is within range of an impulse transmitter.

Wyndham-Clark disintegron rifle
This rifle is really a cellular disrupter based on the blue Treen handguns captured during the XQY incident. Being fatal to any living organism, it is a powerful deterrent against possible aggressors.

TREEN WEAPONS

Space Fleet continues to encounter hostile Treens on a regular basis, and it must be assumed that ISF will continue to do so until hard physical evidence exists to confirm that the Mekon is dead. It is therefore important to have an understanding of Treen hand weapons as well as those developed on Earth.

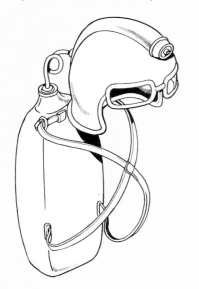

Encapsulation weapon

The Treen encapsulation weapon is an amazing piece of equipment carried by search and snatch troops, in which a special helmet nozzle fires a transparent plastic spray/sheet over the victim. The plastic liquid is carried in a backpack container linked to the helmet, and sets within seconds when exposed to air. The plastic forms a flexible oval bubble around the captive, who doesn't suffocate, as air can pass both ways through the membrane; but they are trapped inside their plastic prison until cut free by special Treen disintegron hand equipment.

Above and right: Treen encapsulation weapon in use.

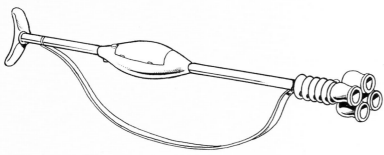

Shoulder flame projector

This favourite Treen weapon draws its ingredients from the air, using a catalyst inside to activate the flame; its range can be doubled by linking two guns together. The projected flame beam is so strong that it can punch through most materials, with the exception of Selektrobot steel, which deflects it.

Cellular disrupter gun

Not to be confused with the red Treen flame pistol that has been adopted by Space Fleet, this blue-coloured gun fires a purple cellular disrupter beam that is fatal to all living things. However, a shield made of lead, doonite, venusium and Mars-sulpher is effective against it.

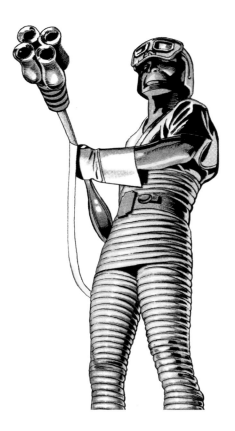

Treens

The Treens inhabit Northern Venus, where they evolved from reptiles. Emotionless, cold-hearted, slaves to science, they are led by a special breed that has a small body and a large brain, called a Mekon. Their ambition is to dominate all other races, but since the Mekon was defeated in 1996 the Treen people has become divided in its loyalties, some embracing peaceful coexistence with other planets while the rest still support the Mekon. The Treens have been a constant source of problems ever since.

The Mekon

A Mekon takes 50 years to mature, and lives for 300 years. He moves around by means of a personal flying chair that has been armoured against all known hand weapons. He is ruthless in his dealings with those who fail him, destroying any who don't deliver, and has proved very elusive when things go against him. A number of times he has been presumed dead, only to resurface at a later date.

Therons

An advanced, peace-loving, brown-skinned, golden-haired race from Southern Venus, led by the elderly President Kalon. They remained isolationist for centuries, but eventually allied with Earth to defeat the Mekon's plans, and have been strong allies ever since.

Atlantine soldier, priest and peasant

Atlantines are not true Venusians, but descendants of people taken from Earth by the Treens 15,000 years ago, for use as slaves. Kept in rural reservations and poorly educated, they became Earth's allies in the Venus war. The skin colour and extra forehead nodule are the result of human adaptation to the Venusian environment.

Mercurians

A friendly, tall, thin but physically very strong race that hates bloodshed. They live in and around Mercury's temperate belt. Their only food source is fay-shaw, a Mercurian vegetable that is deadly to other species. They communicate in a sing-song dialect of vowel sounds. They would rather be left to themselves, and have no interest in travelling off-planet.

Vora

It is believed that Vora, from a distant part of outer space, was the last of his race. Little is known of his background except that he came from a twin-sun system, and always wore a special enclosed environment suit. He was ruthless and power-hungry, and used his Kroopak flying machines to dominate the Thorks. He eventually killed himself rather than be captured.

Thorks

The flamboyant, volatile, ridge-faced Thorks live on the moons of Saturn, where they are born into a caste system based on the moon of their birth – which also determines their size and colour. Under the leadership of Tharl, and with a little help from Earthmen, they succeeded in rebelling against their oppressive upper classes. However, continuing wars of succession mean that the Thorks have since had little to do with other solar system dwellers.

Crypt

An intelligent, peace-loving race from Cryptos, a planet in the system of Los, thought to be about five light years from Earth. They developed an amazing space drive to enable them to come to Earth seeking help to counter the coming Phant invasion. In the end it turned out that it was their diet that made them docile.

Phants

A once warmongering race from Phantos, a rogue planet in the system of Los, who used to enslave the peaceful Crypts. Their dress denoted rank, flamboyance or type of military occupation. It was eventually discovered that their aggressiveness was due to diet, and once this was changed they became as peaceful as the Crypts.

ALIEN IDENTIFICATION

Cosmobes

Cosmobes are tiny aquarian refugees from deep space, who were fleeing their dying I-Cos world and their Pescod enemies. Only one of their fleet of spacecraft stayed on Earth to set up a colony, while the rest travelled on into space. In return they helped Earth fight against the Pescods, and allowed ISF access to their technology.

Pescods

A human-sized aquarian race fleeing from the dying Cosmobe system, but in the process attacking anything in their path with their deadly 'Crimson Death' weapon. Their fleet ended up beneath Earth's oceans, but made the mistake of drilling into a volcano to make an underwater base. All are believed to have been killed in the resulting explosion.

Krevvid

A violent, ten-foot tall, bounty-hunting race from outer space, accidentally released from suspended animation. They were concealed on a red ship found in the Sargasso Sea of Space. Although this crew has perished, more of them could still be somewhere out in space. Exploration teams have been warned to avoid contact with them under any circumstances.

Mystery hostile aliens

An unknown hostile race that arrived in the solar system without warning, took over a mine on a Jovian moon and attacked Earth spacecraft. This incursion was completely wiped out when their mother ship was destroyed in action by a photon blast. Their origins are unknown, and it is unclear whether any more will appear.

Xel

Xel is a Stollite who was stranded on Meit. A very strong, tough, violent and power-crazed alien who got to Earth as a Tempus Frangit stowaway. He tried to seize control on Earth but is believed to have been killed during the Triton invasion crisis. The location of his home world is unknown, so exploration teams need to be vigilant.

Vendals

Ruled by the savage Reshnek, these evil brutes with four arms came from Volk, another world in the Vega system. They had desolated their home planet in wars using a deadly radiation weapon known as the 'Singing Scourge', named from the sound it made when it was operated.

Pittars

Pittars were a colossal, savage alien race that was colonising Jupiter. They killed out of hand anyone who stood in their way, or used them to provide some kind of twisted amusement. However, when they encountered the common cold virus it hit them very hard, as they had no immunity to it, and they fled back into deep space.

Tritons

The Tritons lived in an automated city on Neptune's largest moon. They just ate and slept, which earned them the nickname 'Moonsleepers'. The alien Xel found them, and by a mixture of intimidation and false promises welded them into an army with which to invade Earth. However, they were defeated in space by Earth and Theron forces.

Verans

The Verans are a race of small stature who always wear tough, heavy environment suits that weigh tons – which is not surprising, as they live on the surface of Jupiter, under its huge atmospheric pressure. Since they wear the suits even on Jupiter it seems very likely that at some point they must have lived elsewhere.

Trons

The Trons are a mixture of peaceful races found on the planet Lapri in the Vega system, who were enslaved by the brutal Vendals. It was only the arrival of the Tempus Frangit's crew that tipped the balance and successfully helped the Trons to fight for their freedom.

TREEN CRAFT

TREEN FIGHTER

The original four-fin, single-seat Treen interceptors that operated in the 1996 Earth–Venus war had two types of armament: a nose cannon, and a disabling ray fired from the front of its nacelles, which jammed their opponent's controls.

In 1999 a slightly different four-fin interceptor appeared with the Mekon's fleet on Mercury. This type fought in pairs that could project a linking gas jet between them, and then use that gas field to pass an electrical current through an enemy's ship to stun the crew.

In 2011 a two-seat version appeared, which had a manned gun turret situated above and behind the pilot.

ISF DATA FILE

All four-fin fighters were scrapped after the Reign of the Robots, when strict restrictions were enforced to limit the offensive capability of the Treens.

Left: Two interceptors work together to mount an electrical attack on the Hermes in Mercurian space.

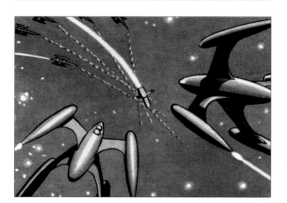

Above: A pilot and gunner version of the interceptor was first encountered in 2011.

Left: 1995 ray-firing interceptors.

TREEN SAUCER

These craft were designed to spy on Earth. They had forward viewing areas but relied more on sensitive scanning equipment to observe and record what was around them. They had the advantage of being able to land and take off vertically, so could alight in very small areas to collect samples or install monitoring equipment. Each saucer carried a crew of nine, consisting of two pilots, a flight commander and six technicians. In addition there was room for essential cargo.

ISF DATA FILE

The Mekon had used saucers to spy on Earth for years, and was going to use them to build a Telezero transmitter on the Moon. However, the attempt was foiled, and with no further use for saucers after the 1996 Earth–Venus war the remaining craft were scrapped. However, a number of them, in various conditions, were subsequently found in the Sargasso Sea of Space, which suggests that they had been used for another mission somewhere that must have gone wrong.

Above and left: Treen saucer front and top views.

EXPERIMENTAL DEEP-SPACE CRAFT

These strange-shaped craft had mixed Treen and Atlantine crews, which meant equipping them for the needs of two races. This was considered very inefficient in Treen eyes, and was not attempted again.

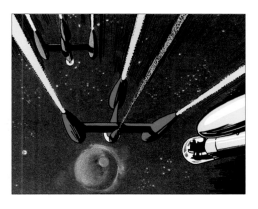

Right: Racing to close on the Red Moon orbiting Earth.

Left: One of the ships fuelling up.

ISF DATA FILE

These craft raced from Venus to aid Earth against the Red Moon, but the experimental weapon they carried only sent the Red Moon careering deeper into the Solar System. They pursued and caught up with it near Mercury, where they destroyed it with a chain reaction bomb.

The Mekon's Telezero ships are designed to link up in space to form a large reflecting mirror that can bounce his Telezero destruction beam around the curvature of planet Earth. Each ship is Telezero-proofed, and has six gun-turrets to defend itself against attack. They also have the novelty of decks that rotated 90° after take-off, when they shift from planetary gravity to artificial ships' gravity.

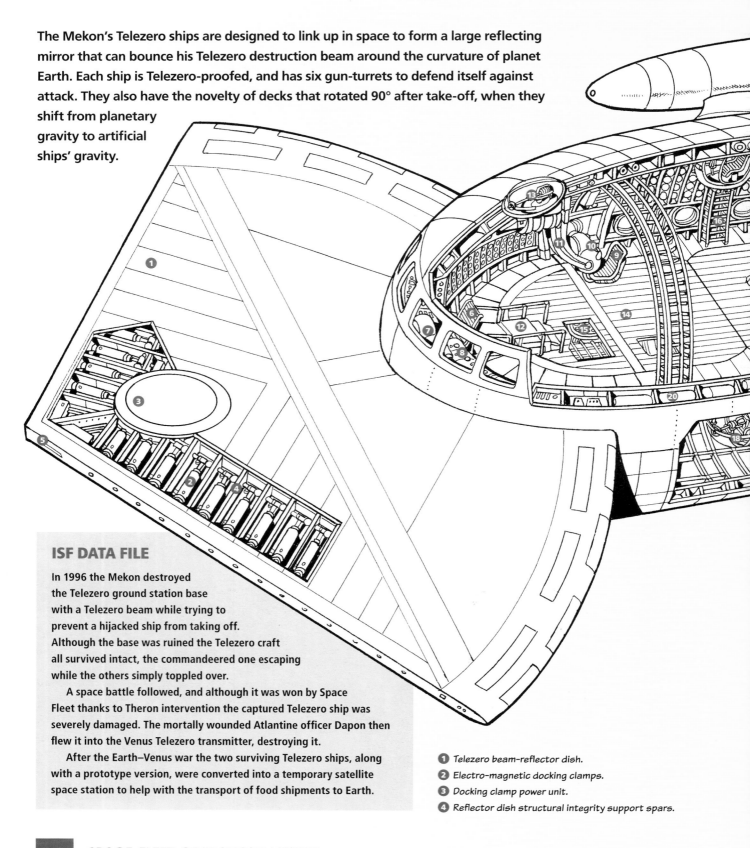

ISF DATA FILE

In 1996 the Mekon destroyed the Telezero ground station base with a Telezero beam while trying to prevent a hijacked ship from taking off. Although the base was ruined the Telezero craft all survived intact, the commandeered one escaping while the others simply toppled over.

A space battle followed, and although it was won by Space Fleet thanks to Theron intervention the captured Telezero ship was severely damaged. The mortally wounded Atlantine officer Dapon then flew it into the Venus Telezero transmitter, destroying it.

After the Earth–Venus war the two surviving Telezero ships, along with a prototype version, were converted into a temporary satellite space station to help with the transport of food shipments to Earth.

❶ Telezero beam–reflector dish.
❷ Electro-magnetic docking clamps.
❸ Docking clamp power unit.
❹ Reflector dish structural integrity support spars.

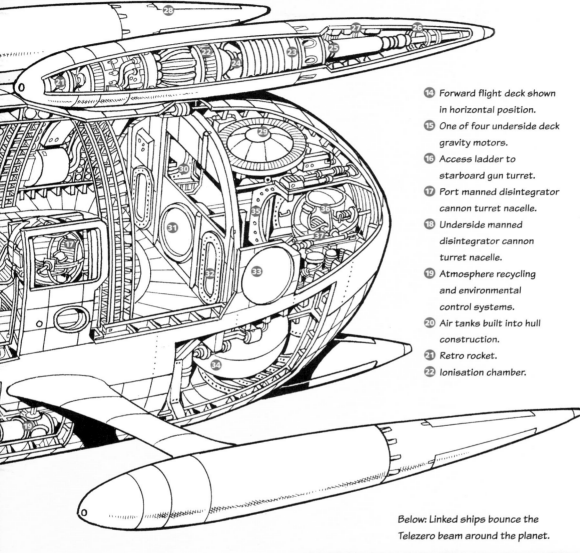

23 Heat sink manifold.

24 Particle accelerator.

25 Treen-originated ion drive particle exhaust pipe.

26 Twin landing legs hold each drive nacelle clear of the ground.

27 Upper drive nacelle landing leg actuators.

28 Drive nacelle landing leg door.

29 Artificial gravity generator controls gravity motors on all decks.

30 Starboard airlock.

31 Access door to rear disintegrator cannon turret nacelle.

32 Port side inner airlock door.

33 Port side outer airlock hatch.

34 Shielded atomic reactor.

35 Turret bulkhead airlock.

36 Rear gun turret in retracted position. In use, the turret bubble canopy is extended clear of the ship's hull to give the cannon operator a 200 degree line of fire. Turret is air tight with integrated life-support system and power unit.

37 Turret deployment rams.

14 Forward flight deck shown in horizontal position.

15 One of four underside deck gravity motors.

16 Access ladder to starboard gun turret.

17 Port manned disintegrator cannon turret nacelle.

18 Underside manned disintegrator cannon turret nacelle.

19 Atmosphere recycling and environmental control systems.

20 Air tanks built into hull construction.

21 Retro rocket.

22 Ionisation chamber.

5 Forward beam-guidance sensor ensures precise manoeuvring of the three Telezero ships during docking procedures.

6 Telezero docking control station.

7 3D video communications transceiver.

8 Docking positioning controls.

9 Pilot's seat shown partly swung down in horizontal flight position.

10 Flight instrument panel.

11 Flight instrument panel positioning arm.

12 Pilot's seating positioning guide rails, used when the craft is in horizontal flight mode.

13 Topside disintegrator cannon, controlled from pilot's instrument panel.

Below: Linked ships bounce the Telezero beam around the planet.

OVERVIEW

The Anastasia was designed and built in 1997 by Sondar, using both Treen and Theron technology, as a special thank you gift to Dan Dare for freeing northern Venus from the Mekon's grip the previous year. The two-seater was named after Digby's Aunt, to honour her for foiling the Mekon's plans by spotting a coded message left by Digby.

It incorporates four different propulsion methods: standard impulse motors for space flight; jets for flying as an aeroplane (with the wings out at right-angles); Theron magnetic motors that enable the craft to hover at any height in any gravitational field; and rocket-fuel engines that can be used as take-off boosters (with the wings swept back) or when beyond impulse power range.

In addition, once on the ground the wing-and-engine section can be parked, while by pulling two levers the front section separates for use as a gyro car.

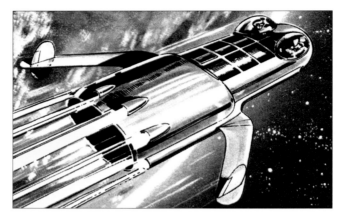

Left and below: Wings extending for horizontal flight.

Bottom left: Space tube launch from the Galactic Galleon, with wings folded.

ISF DATA FILE

The Anastasia has been a major factor in a number of historic events, including saving the Mars evacuees; landing and rescuing Professor Peabody from the Red Moon, and towing the chlorophyll lamp that lured the Red Moon away from Earth; enabling Colonel Dare to reach Venus unobserved during the Robot occupation; participating in the Phoenix Mission; working in space during the Cosmobe communication black-out; undertaking several missions on Terra Nova; assisting people in the super Treen war; and serving during the Mushroom crisis. It was even used by Xel as a command ship during the Triton invasion.

During all of this she has endured a good many scrapes, being damaged by atomic bomb debris near the Red Moon and crash-landing at SFHQ in 1999; being lost in space the same year, then found, serviced, and later shot up in Venus orbit in 2011; crashing on and being lifted from the Venus lava plains soon afterwards; being damaged by flying debris in the Sargasso Sea of Space in 2012; being hog-tied to a Cluster ship by the Cosmobes in 2012; stolen on Venus in 2013; involved in fighting Nagrabs on Terra Nova, and while still there in a war against Gax in 2013; resurrected from mothballing after the Zylbat was destroyed; rescuing civilians from a volcano on Venus in the Super Treen war; damaged again in the Mushroom crisis; and shot down and crashing in Antarctica in 2020. She has now been rebuilt for the fifth time.

*Above: Anastasia
as a ground car.*

*Below: Refugees climbing
aboard the Anastasia.*

GYRO-CAR DEPLOYMENT

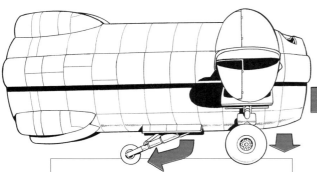

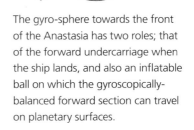

Deployed only when the Anastasia's gryro-car section is required for planetary surface exploration, the rear stabilisation strut supports the rear of the ship to prevent it from overbalancing. Once the gyro-car has re-attached, the strut retracts and the Anastasia is ready for take-off.

The gyro-sphere towards the front of the Anastasia has two roles; that of the forward undercarriage when the ship lands, and also an inflatable ball on which the gyroscopically-balanced forward section can travel on planetary surfaces.

THE ANASTASIA

Revolutionary in both function and design, the Anastasia is the most famous spaceship used by the ISF, almost exclusively by Dan Dare himself to whom it was given by Sondar. When constructed, it represented the ultimate in compact technological design, boasting four propulsion systems unheard of in a craft of this size. Although rebuilt and altered a number of times over the years, it remains Space fleet's most iconic spacecraft, despite its alien origins.

28 Telescopic swing-wing sweep-back cover.
29 Airlock control panel.
30 Fuel injection pumps.
31 Theron-designed magnetic motors.
32 Main fuel tank for jet motors.
33 Rocket fuel tanks containing Space fleet blended liquid fuel.
34 Electrosphere primary accelerator.
35 Impulse wave rectifier.
36 Main motor: Space fleet impulse wave system.
37 Electrostatic chamber: secondary accelerator.
38 Electrostatic propellant tank.

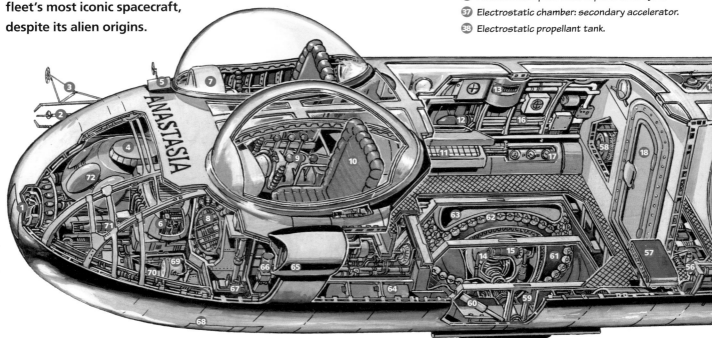

1 Space fleet Maintenance Vehicle towing hook.
2 Particle beam gun.
3 Particle Beam Weapon visual target and range finder array (deployed).
4 Gryocar propulsion fuel balance feeder.
5 Particle Beam Weapon target array (deployed).
6 Gyrocar propulsion power mixer.
7 'Control A' flight instrument panel.
8 Atmosphere data computer.
9 Atmosphere flight controls.
10 Co-pilot's seat.
11 Tip-up plotting table with padded cover which serves as a seat.
12 Starboard Gyrocar ground propulsion turbo jet.

13 Air-circulation vent.
14 Gyroscope maintains vehicle's balance in Gyrocar mode.
15 Gyrosphere and air cushion air pump.
16 Control conduits to main motors.
17 Life-support controls.
18 Air-sealed door to first airlock compartment.
19 Airlock pressure indicators.
20 Air-sealed diaphragm door to main airlock.
21 Rear section airlock door.
22 Water tank and recycling/purification system.
23 Airtight washroom door.
24 Toilet (can be adapted for zero-gravity use).
25 Automatic washer and dryer.
26 Airtight storage room door.
27 Topside access hatch.

39 Rear stabilisation strut, incorporating landing wheel. This is located forward of the central lower rocket motor and is never used during normal landing/take-off situations. It is deployed, however, when the Anastasia has already landed and the Gyrocar section needs to be used for surface exploration. Without it, the ship's rear would topple backwards.
40 One of six rocket motors.
41 Primary exhaust nozzles.
42 Port jet motor.
43 Electricity generators.
44 Rudder servo system actuators allow the rudder to face forward when the swing wings retract aftwards. In atmospheric flight the rudder can rotate to aid steering.

45 Rudder servo system.

46 Aileron control actuators.

47 Impulse wave condenser cylinder.

48 Port landing gear hatch.

49 Port landing wheel.

50 Synchronised tension spars operate the swing wing systems.

58 Life-support system incorporating liquid oxygen converter and zyolithic crystals to prevent condensation, along with algae to siphon carbon dioxide out of the cabin.

59 Gyrosphere axle vertical support and suspension stanchions.

60 Hydraulic gyrosphere axle suspension.

62 Low friction independently suspended ball bearing layer around the top half of the landing sphere allows smooth operation of the inflated gyrosphere when inflated for landing, take-off and Gyrocar travel functions.

63 Landing gear sphere shock-absorbing inflatable air cushion.

64 Gyrocar fuel tanks.

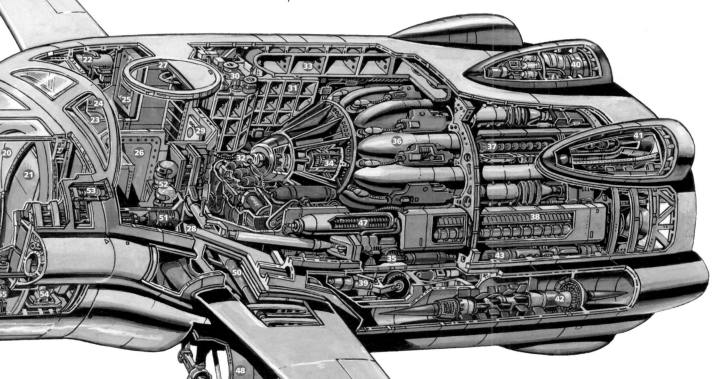

51 Hydraulic pump and fuel conduit shut-off valves, used when the Gyrocar separates from the rear of the craft.

52 Spacesuits, tools and weapons storage.

53 One of eight electro-magnetic docking clamps.

54 Aft port Gyrocar stabilisation landing strut, deployed when the Gyrocar is stationary.

55 Aft port Gyrocar stabilisation jet.

56 Underside multi-function probe winch.

57 Cabin seating.

61 Forward landing gear and gyrosphere. In flight mode the gyrosphere is partially deflated; when used for landing, the sphere inflates around the central Gyroscope core, opening the landing bay doors.

65 Jet intake and braking rocket thruster.

66 Forward port stabilisation jet, one of four used in conjunction with the gyroscopically-balanced sphere to maintain the Gyrocar's balance and provide additional manoeuvrability.

67 Forward port Gyrocar stabilisation strut, used when the Gyrocar is stationary to prevent it from tipping over.

68 Stabilisation jet and strut hatches.

69 Searchlight (retracted).

70 Searchlight hatch.

71 Forward missile launch tube.

72 Zadix 2 computer.

THE ANASTASIA

The Anastasia's dual-control systems enable the craft to be flown effectively by both cockpit occupants if the pilot or co-pilot is engaged in other tasks or is absent. This eliminates the need for seating positions to be exchanged during flights or if the ship is engaged in battle. Although the instrument panels control different functions, each can be seen from both seats easily, while the control columns' functions are identical and instantly switchable between the two.

Left:
Towing the
chlorophyll
light.

1 Cabin seating.
2 Fuselage construction made from Fransite bonded with ceramic additives providing radiation and re-entry burn-up protections.
3 Co-pilot's seat.
4 Toughened laminate polarised canopy, bonded with heat resistant ceramic and Fransite additives to protect pilots from solar glare and radiation. Also resistant to re-entry burn-up.
5 Control B instrument console with multi-function screen displaying radar and navigation data.
6 3D Video communications screen can also display avionics data from Zadix II computer.
7 Communication console and speaker.
8 Port side control column; can be pushed forward under console to allow easier access to cockpit seat.
9 Dual-mode floor pedals control jets in flight, and vehicle functions in 'Gyro Car' mode.
10 Atmosphere flight controls operate swing wings, ailerons and tail fins.

11 Computer function monitor lights.
12 Pilot's seat.
13 Control A instrument console: dials and switches operate and display gravity, heating, light, acceleration and attitude where appropriate, plus gravity-linked inertial reduction control functions.
14 Radio microphone.
15 Landing undercarriage controls.
16 Stabilisation jet maintenance access panel.
17 Forward starboard jet intake cowling.
18 Impulse motor activation control.
19 Cockpit pressurised bulkhead.
20 Gyrocar propulsion power mixer.
21 Gyrocar propulsion fuel balance feeder.
22 Atmosphere and external environment data processor.

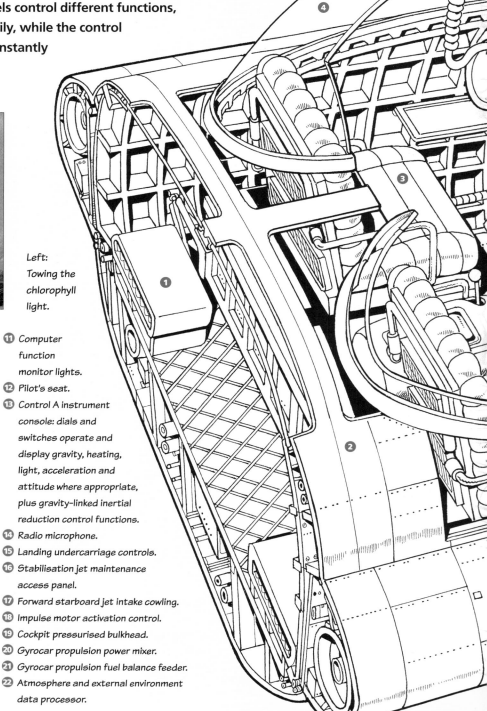

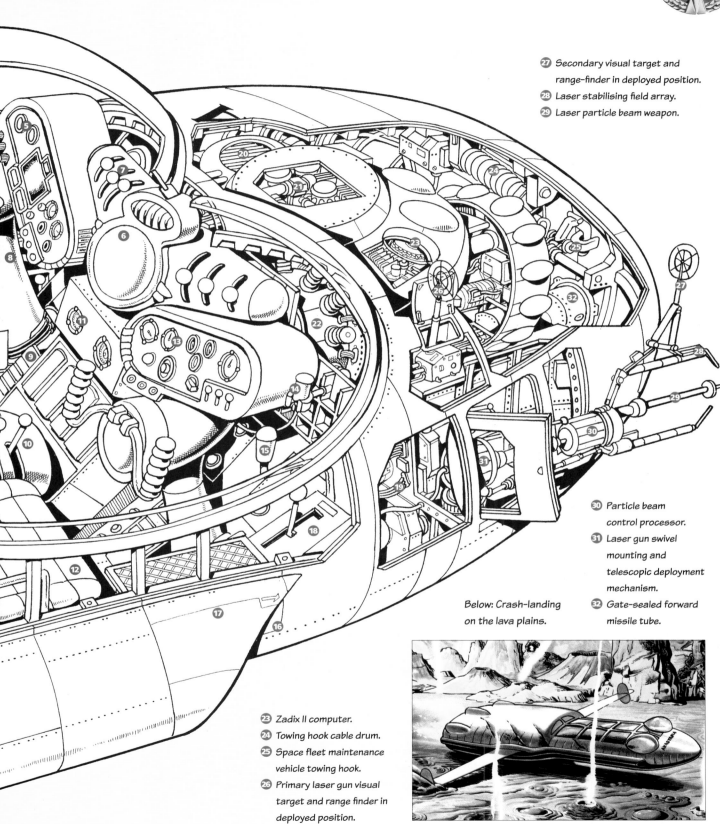

27 Secondary visual target and range-finder in deployed position.
28 Laser stabilising field array.
29 Laser particle beam weapon.

30 Particle beam control processor.
31 Laser gun swivel mounting and telescopic deployment mechanism.
32 Gate-sealed forward missile tube.

Below: Crash-landing on the lava plains.

23 Zadix II computer.
24 Towing hook cable drum.
25 Space fleet maintenance vehicle towing hook.
26 Primary laser gun visual target and range finder in deployed position.

TREEN BATTLE CRUISER

Built during the Mekon's occupation of Mercury, these craft were armed with X-bombs and four nose cannon. They normally carried a crew of nine, and could stay in space for prolonged periods. The flight nacelles were mounted on retractable fins, enabling the craft to enter space station docking tubes. With the aid of four retractable support legs it could also make tail-first planet landings.

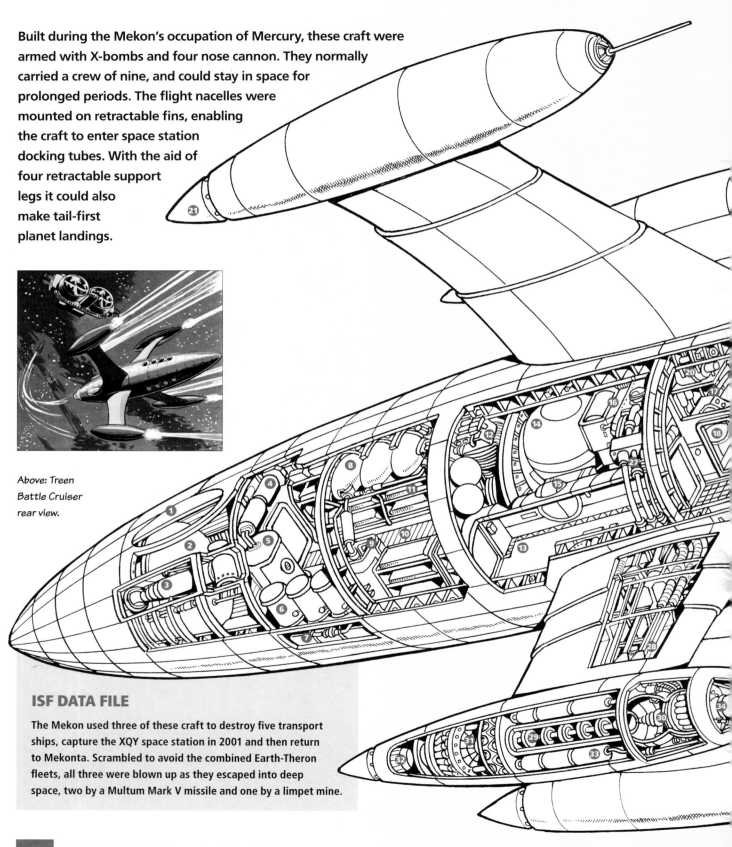

Above: Treen Battle Cruiser rear view.

ISF DATA FILE

The Mekon used three of these craft to destroy five transport ships, capture the XQY space station in 2001 and then return to Mekonta. Scrambled to avoid the combined Earth-Theron fleets, all three were blown up as they escaped into deep space, two by a Multum Mark V missile and one by a limpet mine.

1. Airlock outer hatch.
2. Gravity adjustment generator enables crew entering the airlock to be re-orientated with the ship's own gravity direction when landed vertically.
3. One of four landing struts.
4. Water tank.
5. Personal cleansing station.
6. Treen nutrient tanks.
7. Nutrient storage and mixer tank.
8. Compressed air tanks.
9. Crew waste management cubicle.
10. Crew quarters.
11. Port side crew bunks.

Above: Treen Battle Cruiser with bomb doors open.

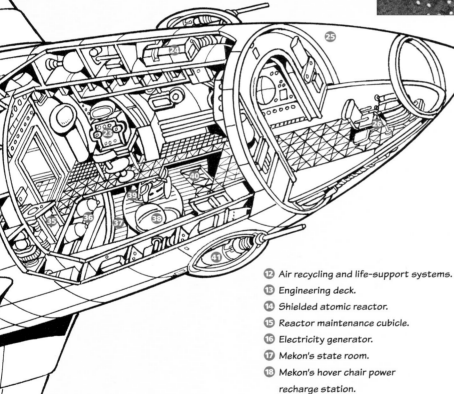

12. Air recycling and life-support systems.
13. Engineering deck.
14. Shielded atomic reactor.
15. Reactor maintenance cubicle.
16. Electricity generator.
17. Mekon's state room.
18. Mekon's hover chair power recharge station.
19. Nutrient supply tank.
20. Mekon's personal air supply, recycling and life support system.
21. Drive nacelle shown in extended position for flight; nacelles can retract if ship needs to dock within a small space or landing area.
22. Multifunction antennae provide range-finding data for disintegrator cannons, communications and external environment monitoring.

23. Communications console.
24. Disintegrator gun power unit.
25. Polarised and heat-resistant flight deck canopy.
26. Pilot's seat and console.
27. Ion drive power conduits.
28. Wing deployment motors allow drive nacelles to retract when entering small airlocks or spacecraft hangars.
29. Ion particle accelerator.
30. Particle gun.
31. Ion Drive gate seal, used when take-off rocket is in operation.
32. Take off rocket.
33. Rocket fuel tanks.
34. Manoeuvring/retro rocket.
35. Starboard flight deck consoles control navigation and armament systems.
36. Racked bombs.
37. Bomb bay hatch.
38. Solar cell orbital space prison, with 12-hour air supply and life-support system.
39. Solar cell transit clamps.
40. Access hatch connecting bomb bay with flight deck.
41. Automated disintegrator guns; one of four operated from the flight deck.

TREEN CRAFT

Two other types of Treen deep-space craft entered service during the robot occupation, and both remain in use to this day. The first is a large, non-streamlined cruiser designed for deep-space use only. It normally travels only between space stations, but can dock with other craft when necessary and is equally at home transporting passengers or cargo.

The second is capable of planetary landings as well, and has an outer fuel tank built around it that gives it much-extended range. This fuel tank can be refuelled or jettisoned as circumstances require.

Both craft carry in-space concertina docking tubes as standard.

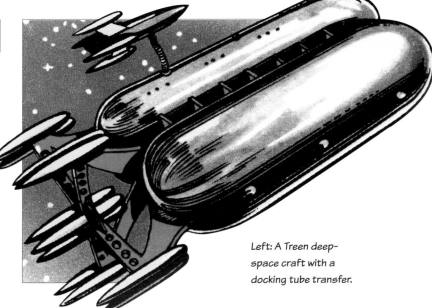

Left: A Treen deep-space craft with a docking tube transfer.

ISF DATA FILE

In 2011 Colonel Dare escaped the Mekon's clutches in a double-tank craft, and the same ship was later used by Captain King and Angus MacFarlane to decoy Treen forces away from his activities.

Middle: Deep-space craft with an outer fuel tank and jettisoning the outer tank (right).

Left: Making a landing

MEKON'S COMMAND SHIP

As used in the robot invasion and throughout the occupation of Earth, the Mekon's command ship was a fast interplanetary craft that was equally at home underwater and in deep space. It could make precision vertical landings on its underbelly, and if necessary could be controlled by just one person, or even operated by remote control.

ISF DATA FILE

The Mekon's command ship disappeared in 2011 after the robot occupation was defeated, only to reappear years later when the Wandering World crossed the outer reaches of the Solar System. At that time the Mekon was its only occupant.

'SOLID SPACE' CRAFT

When the Mekon resurfaced during the 'Solid Space' crisis in 2016 he had a number of new spacecraft with him. It is believed that all were destroyed, but it is important that they are recognised should they reappear in the future.

Below: The Mekon's headquarters with the 'Solid Space' satellite in the foreground.

Above: The Mekon's command ship

Left: The satellite that created 'Solid Space'.

Below: The Mekon's main ship.

Bottom: The fast, four-man scout ship.

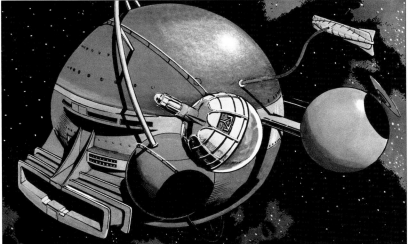

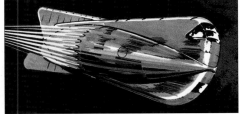

Originally built to transport food supplies to Earth, these unarmed heavy-duty carrier ships can lift off direct from the surface of Venus, which avoids the need to tranship goods at M.E.K.1 and consequently saves valuable time.

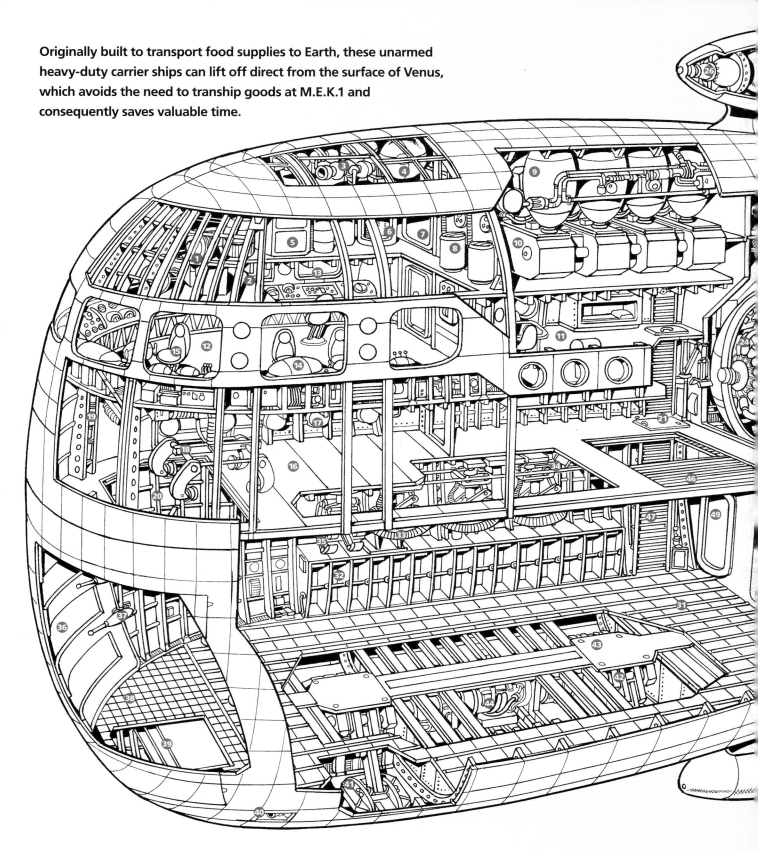

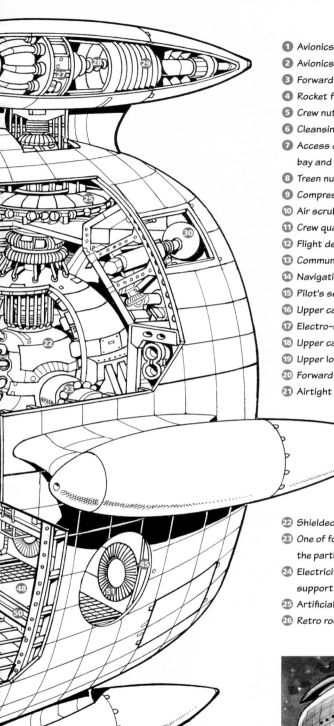

1. Avionics and Communications Bay.
2. Avionics Bulkhead.
3. Forward upper port retro rocket.
4. Rocket fuel tanks.
5. Crew nutrient storage tank.
6. Cleansing cubicle.
7. Access door to life-support systems bay and crew quarters below.
8. Treen nutrient baths.
9. Compressed air tanks.
10. Air scrubbing/recycling processors.
11. Crew quarters.
12. Flight deck.
13. Communications console and video screen.
14. Navigation station.
15. Pilot's seat.
16. Upper cargo bay.
17. Electro-magnetic cargo transit clamps.
18. Upper cargo bay handling grab.
19. Upper loading bay grab control hydraulics.
20. Forward loading hatch to upper cargo bay.
21. Airtight access ladder hatch.

22. Shielded atomic reactor.
23. One of four electricity generators powering the particle accelerators in each nacelle.
24. Electricity generator powers gravity, life support and ship's systems.
25. Artificial gravity generator.
26. Retro rocket.

27. Ionisation chamber.
28. Treen-originated particle accelerator.
29. Heat sink manifold.
30. One of four rear landing struts enabling the carrier to land on its tail as well as horizontally.
31. Lower cargo bay.
32. Wall- and floor-secured transit containers used to carry smaller items of cargo including robots, weapons and tools.
33. Gravity-assisted cargo handling clamps.
34. Transit clamp positioning rail.
35. Cargo handling grab.
36. Starboard cargo bay door.
37. Bay door hydraulics.
38. Forward hinged section of cargo deck enables loading ramp to be deployed.
39. Cargo bay loading ramp.
40. Forward lower port retro rocket.
41. Forward port landing strut. The carrier's dual landing configuration enables cargo to be loaded through the doors at the front.
42. Rear port landing strut hydraulic rams.
43. Landing gear support stanchions.
44. Cargo bay gravity power generator.
45. Port airlock outer door.
46. Rear loading hatch to upper cargo bay.
47. Starboard access ladder to all decks.
48. Rear airlock.
49. Airlock inner door.
50. Rear lower cargo bay door provides loading access from rear gravity-controlled airlock.
51. Gravity-adjustment generator enables crew and cargo entering the rear airlock to be re-orientated with the ship's own gravity direction if it has landed vertically.

Right: Delivering the Treen Space-Lab to M.E.K.I.

ISF DATA FILE

These craft were used by the Mekon to carry his invasion robots to Earth in both 2002 and 2011.

In 2012 one brought the refurbished Treen Space-Lab up to M.E.K.1 for use by the second Sargasso Sea of Space expedition.

SPACE SHARKS

These single-seat, rocket-powered fighters are armed with four nose cannon plus two missile launchers in the front of their rocket nacelles. They usually operate in squadrons of three or six.

Above: Space Sharks attacking Z-19s.

ISF DATA FILE

Red and yellow Space Sharks were first seen clearing the way for the Mekon's robot invasion fleet in 2002. They also acted as escort to the second Earth assault fleet (of Selektrobots) in 2011. However, their main purpose was to patrol Venusian space from M.E.K.1. In 2011 eight attacked the Anastasia as she was trying to reach Southern Venus. Four were shot down by Space Fleet personnel manning Treen fighters and the others were driven off, but the Anastasia was damaged and crash-landed on the lava plains of Venus.

After the reign of the robots, the Space Sharks appeared in new colours – green and yellow. For a time the Treens were allowed only one squadron, named the Green Magnets, which was under the direct control of Governor Sondar.

In 2012 the Green Magnets were called into action against the Pescod fleet, with disastrous results. Half of the squadron was destroyed by the 'Crimson Death' weapon, and the rest retreated without inflicting any damage on their opponents.

1. One of four nose-cone mounted disintegrator cannons in deployed position. Each can retract into nose cone during atmospheric re-entry procedures.
2. Cannon deployment arm.
3. Disintegrator cannon retraction guide tracks.
4. Disintegrator cannon power and charging unit.
5. Forward retracted telescopic landing leg.
6. Landing leg door.
7. Instrument panel shroud.
8. Polarised heat-resistant canopy.
9. Control column.
10. Vectoring rocket control pedals.
11. Control console and engine throttle.
12. Air cleansing/recycling unit removes carbon dioxide from cockpit.

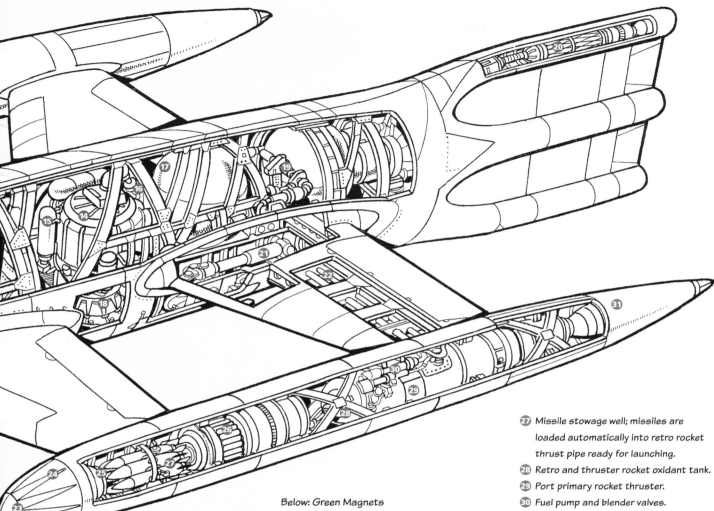

27 Missile stowage well; missiles are
    loaded automatically into retro rocket
    thrust pipe ready for launching.
28 Retro and thruster rocket oxidant tank.
29 Port primary rocket thruster.
30 Fuel pump and blender valves.
31 Thrust nozzle vectoring vanes.

*Below: Green Magnets
attacking the Pescod fleet.*

13 Compressed air tank.
14 VTOL rocket thrusters.
15 VTOL oxidant tank.
16 VTOL fuel pump.
17 Rocket fuel tanks.
18 VTOL thruster nozzle.
19 Fuel distribution valves.
20 Tailfin-mounted additional
    booster rockets.
21 Rear port retracted telescopic landing leg.
22 Fuel feed conduits.
23 Missile hatch/retro rocket nozzle door.
24 Retro rocket vectoring vanes.
25 Rocket thrust pipe doubles as missile
    launch tube when not in use.
26 Port retro rocket.

THERON CRAFT

THERON FIGHTER

These small, very fast and manoeuvrable, single prone pilot fighters have a cluster of eight guns in the nose of the craft, which can deliver a concentrated burst of fire over a wide area of the target. Coupled with a tight turning circle, and a very impressive speed, they are extremely hard to catch and, despite their size, are not to be underestimated.

ISF DATA FILE

These craft fought in the first ever space battle in 1996, in which they successfully defended a severly damaged Telezero ship. They ran rings around the Teen four-fin fighters that were attacking the Telezero ship, destroying many of them in single bursts of fire. They were called into action again in 2001 when the Mekon returned to Mekonta. This time they joined forces with Earth's Elite Squadron, but on arriving at Mekonta found the Mekon and his ships had already departed. A year later they were all destroyed in their hangers in a sudden attack by the Mekon's Selektrobots.

Above and left: Theron fighter craft in action.

COMMAND CRAFT

The two-deck command craft is even more impressive. As well as matching the smaller fighters in speed, it has four upper-deck multi-gun turrets plus a single mid-ship lower turret. It can link up with other craft in space by means of a transfer tube attached to its airlock.

It usually flies with a crew of 20: a commander, pilot, co-pilot, communications officer, medical officer, five crew manning the guns, the others acting as reserve crew and ship boarding parties.

Above: Command craft with guns deployed.

Left: Command craft and fighters.

DEEP-SPACE CRAFT

These long-established deep-space craft known as Saurons are a favourite of the Therons, so much so that more were built after the 2002–11 robot occupation that destroyed the previous fleet. Surprisingly, the Therons have released very little information about them – including their range, top speed and the full extent of their armament – but they are known to carry very powerful missiles, can at least match the XC-9's speed, and have a crew of around 20. They also don't need launch ramps, as they can land or take-off just like an aircraft, because their wings can be extended or retracted.

Above: Theron deep-space craft.

ISF DATA FILE

It was a Sauron craft that brought the original Earth/Venus expedition home in 1996, and ten of them also assisted in defeating Xel's Triton in fleet in 2020.

Tharl's formidable flagship has four space propulsion units equally spaced around an external ring, mounted on top of four paired heavy-calibre space cannon that can elevate or swing sideways as required. There are also two fixed forward-facing guns near the ship's nose, and Katabolic gun batteries along the port and starboard sides to defend against Kroopak attack. The rear is guarded by four shielded gun turrets.

Right: Front close-up.

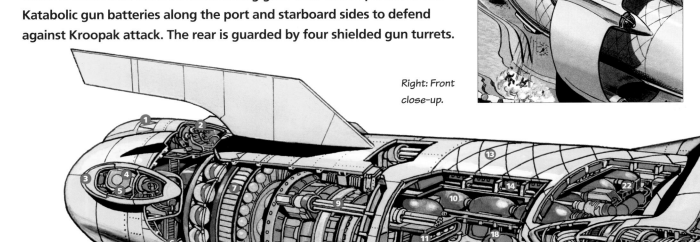

1. Disintegrator cannon nacelle.
2. Airtight gunner's cockpit.
3. Nacelle cannon hatch in closed position.
4. Aft starboard twin-barrel disintegrator cannon.
5. Armoured gun battery nacelle.
6. Ladders from access corridor leading to aft gun batteries.
7. Gravity generator provides internal ships gravity and power from the reactor to energise the repulsion gravity band on the command ring.
8. Atomic reactor provides power for life support, gravity and main engines.
9. Power conduits.
10. Compressed air tanks.
11. Reactor and engine control console.
12. Engine control room.
13. Main hull gravity repulsion panels.
14. Gravity repulsion power units work in conjunction with command ring air scoop to stabilise the ship in hover mode.

15. Rear starboard landing leg (retracted).
16. Rear exit provides access to power plant if repairs are needed and replacement components are required. Also provides access to under-reactor corridor leading to aft gun batteries.
17. Access lift to engine control room.
18. Suction clamp control console: clamp can also be operated from the bridge.
19. Electro-magnetic suction clamp in retracted position. Clamp can be repositioned to forward end of hold. Clamp is used to enable small spacecraft to dock or to load cargo.
20. Flexible clamp head can grip any surface.
21. Storage hold and docking bay.
22. Atmosphere cleansing, filtering and recycling systems.
23. Command ring.
24. Communications antenna and long range radio beam transceiver.

25. External environmental sensors.
26. Command bridge.
27. Thork-originated ion drive, powered by radial impulse generators.
28. Magnetic airscoop gravity repulsion band allows ship to hover.
29. Airscoop exhaust vanes.
30. Airscoop power cells.
31. Magnetic airscoop brakes.
32. Command ring circular access corridor.
33. Radial impulse generator.
34. Four armoured main gun battery turrets, each containing twin disintegrator cannons.
35. Command ring armour plating.
36. Access to main gun turrets.
37. Lift to all decks.

38 Starboard airlock.

39 Space telesender teleportation tubes.

40 Starboard machine cannon, hand-operated.

41 Cannon hatch, used in atmosphere only.

42 Infiltration-agent's skin colouration mixer tanks.

43 Field agent's skin colouration spray pool changes an operative's body colour for infiltration missions.

44 VTOL fuel and blender tanks.

45 Forward starboard VTOL thrusters.

46 Forward starboard landing leg (retracted).

47 Suction nozzle gravity compensation processor.

48 Suction nozzle vents.

49 Suction nozzle used to collect ship's supplies: suction power is variable, using gravity compensation processors to ensure there are no loading mishaps.

50 Suction nozzle cargo loading pad.

51 Gun hatch protects weapons from micro-particles or re-entry heat when not in use.

52 Starboard 'katabolic' gun batteries, one of five operated by two gunners.

53 Starboard cannon operator's bay.

54 Armoured cannon operator's protective bulkhead.

55 Port cannon control and operator's bay.

56 Language learning memorising machine, used by infiltration agents.

57 Stereo viewer communications and visual display unit.

58 Forward control room.

59 Forward scanner display screen.

60 Multi-role navigation display screen.

61 Port cannon particle beam control processor.

62 Forward starboard disintegrator cannon operated from bridge or forward control room.

63 Forward manual control cockpit. Used if the control centre on the ring bridge is out of action.

64 Pilot's couch.

65 Co-pilot's seat.

66 Scanner data processing unit.

67 Forward scanner dish.

68 Power cells for 'katabolic' guns.

69 Double doors to storage hold and docking bay shown in closed position

ISF DATA FILE

When Tharl was a rebel leader he had two identical craft built to confuse the Rootha regarding his whereabouts. One was destroyed in the battle for Titan in 2000, and the other disappeared soon afterwards, only to reappear in the Sargasso Sea of Space years later. In 2012 it was restored there, and after helping to defeat the Mekon it returned to Saturnian space.

COMMAND AND CONTROL CENTRES

There are two command and control centres – one in the external ring, the other in the nose – a large hold able to accommodate smaller spacecraft, which are taken aboard with the aid of a magnetic grapnel, and a retractable nozzle able to suck in extra supplies when the ship flies close to planetary surfaces. In addition it has – unique to Tharl's fleet – an in-space teleportation system that facilitates ship-to-ship personnel transfers.

KROOPAKS

Kroopaks or 'Black Cats' are remote-controlled flying machines capable of operating both in space and within planetary atmospheres. Designed by Vora to intimidate and control the population of Numidol, they are armed with four prongs that can cut through metal, a 'green smoke' that vaporises most materials and paralyses living creatures, and a white, spiralling 'zesto' heat ray that can melt almost anything. They can transmit images and sound back to their operators, have an electric brain that can be remotely programmed, and are protected by cosmic ray armour. However, they are vulnerable to a direct heavy artillery hit or Katabolic bombardment, the latter overloading their electric brain, which then explodes. Between planets or moons, Kroopaks are transported in space hives that can carry a thousand at a time. These hives can either be piloted manually or operated by remote control from a manned ship.

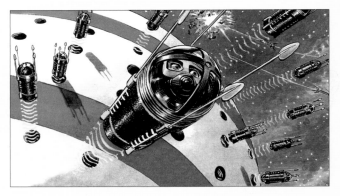

Top: 'Black Cats' emerging from a hive.

Above: 'Black Cats' rear view

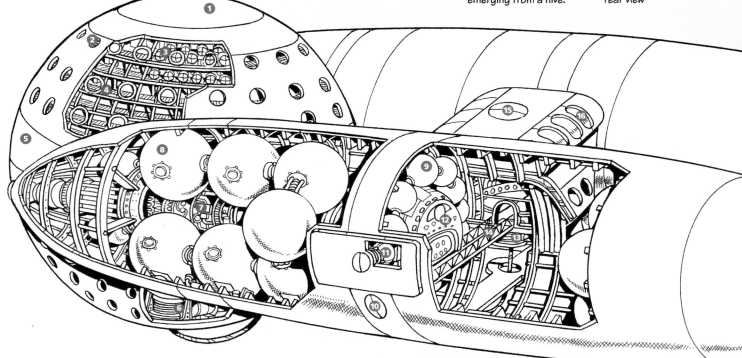

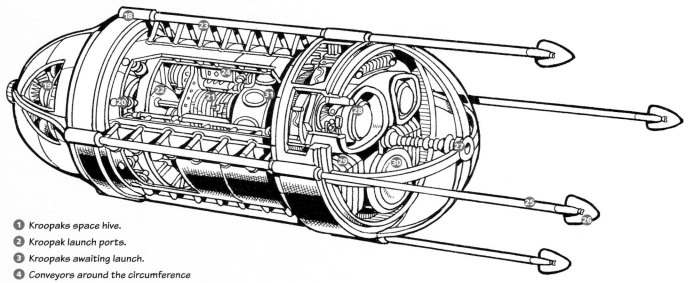

1. Kroopaks space hive.
2. Kroopak launch ports.
3. Kroopaks awaiting launch.
4. Conveyors around the circumference of each storage deck deploy each Kroopak at precise intervals.
5. Hive mid-section Kroopak storage decks.
6. Hive towing cable drum deployment/rewind motor.
7. Starboard nacelle chemical rocket.
8. Fuel tanks.
9. Compressed air tanks supply life-support systems and provide an oxidant to fuel blenders for rocket motor.
10. Engineering deck outer airlock hatch.
11. Starboard manoeuvring thruster.

12. Transporter engineering and life-support control consoles.
13. Transporter cable drum deployment/rewind motor.
14. Engineering deck gantry.
15. Topside airlock to crew quarters.
16. Hive transporter control room.
17. Retro rocket.
18. Aft-facing cameras.
19. Engine exhaust nozzle.
20. Variable pulsed pitch and yaw jets provide precise Kroopak steering capabilities and allow the craft to rotate at high speed, so that the forward-mounted blades can operate as a drill.
21. Turbine drive motor.
22. Combustion chamber.
23. Spiral ray energy processors.
24. Spiral ray energy converter powered by adjacent turbine.

25. Spiral ray targeted field localisers.
26. High density blade cuts through all known metals; each can be used individually or all four can be used to cut larger holes if the craft rotates at high speed.
27. Green paralysing gas emitter.
28. Forward mounted stereoscopic cameras relay images to crew on the transporter or elsewhere if required.
29. Electronic brain.
30. External environment sensor is used in conjunction with the cameras to determine the speed, range and composition of any given target. System also measures external temperature, pressure, radiation in both vacuum and atmospheric conditions.
31. Solid fuel chamber and rechargeable power core.

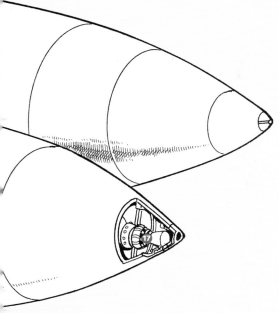

ISF DATA FILE

In the battle for Titan in 2000 the 'Black Cats' were tricked into attacking a doomed ship, which enabled Tharl's rebels to win.

At the same time Vora took a fleet of 2,000 space hives to invade Earth. These were ordered to self-destruct in the nick of time, but at least one hive malfunctioned and ended up in the Sargasso Sea of Space. In 2012 this hive was activated to protect the Phoenix Mission, but was later destroyed.

The Crypts presented this as a faster-than-light vehicle when they reached Earth, but in truth it actually has two propulsion systems – rockets for take-off, landing and near-planetary manoeuvres, and Tengam drive for interstellar travel – and could in fact only attain a speed just below the speed of light.

A Air recycling unit.
B Air tank.
C Steering controls.
D Entry hatch.
E Pilot/personnel couch.
F Pre-deployment rotor storage well.
G Rotor electric motor and battery.
H Fully deployed telescopic rotors steer and reduce the speed of the pod's descent.
I Rocket motor provides limited distance travel from the ship.
J Disintegrator Cannon, installed by ISF.
K Instrument panel.
L Overhead steering control rods.
M Computerised Steering Rod systems.
N Couch angle adjustment spars.
O Entry hatch.
P Foot rest.
Q Life support and engine maintenance panel.
R Life support and atmosphere recycling unit.
S Rocket thruster attitude correction track
T Fuel tanks.
U Twin Rocket thrusters swivel 360 degrees for precise steering and encircle engine fuel nacelle for attitude correction manoeuvering.

ESCAPE POD

For emergency planet-fall there are torpedo-shaped, single-occupant escape craft for the entire crew. These automatically home in on the nearest planet, using rear-mounted helicopter blades to make a safe landing.

EXTERNAL INSPECTION POD

These craft also carry a small single-man spacecraft in which the pilot lies prone, to carry out hull inspections in space.

1 Energy field localiser array.
2 Tengam ray projection unit creates a force field buffer zone ahead of the ship to eliminate damage from space debris. The unit can also provide force field protection in the event of an attack around the vessel.
3 Pilot's instrument console.
4 Pilot's seat.
5 Navigation console.
6 Forward flight deck gravity plating generator shroud.
7 Starboard toughened laminate transparent gun turret, polarised, radiation proofed and heat resistant.
8 Access hatch leading to underside ISF-installed retractable gun turret.
9 Inter star craft computer and communications display screen.
10 Tengam drive visual display screen.
11 Drive control operating systems.
12 Tengam drive power distribution conduits.
13 Tengam drive magnetically shielded housing with rear-facing observation window.
14 Los Crystal – named after the sun from which it was extracted – provides power for the Tengam drive supplied by three magnetic elements taken from three worlds in the Crypt's home star system.
15 Los Crystal power take-off coil.

16 Air-recycling and life-support systems operate vessel's air condition systems plus the suspa-cell suspended animation units.

17 Water purification and recycling plant.

ISF DATA FILE

Tengam drive is constructed from magnetic elements drawn from Los (a yellow star), the planets of Cryptos and Phantos, and the moon of Nula, all in the Los system. When in operation, in order to be drawn to or repelled from its home system it projects a ray field around the craft through which nothing solid can pass, while the crew safely hibernate in 'suspa-cells' (suspended animation units) when necessary.

23 Stellascope long-range visual scanner in deployed position.

24 Stellascope hatch.

25 Transparent armoured bubble behind hatch door prevents atmosphere loss when stellascope is deployed.

26 Personal hygiene station.

27 Observation deck.

28 Observation deck gravity plating generator shroud.

29 Additional life-support systems.

30 Power conduits linking Tengam drive power core to main engines.

31 Engineering deck.

32 Tengam drive power conduits.

33 External inspection pod, primarily used to carry out repairs or maintenance.

34 One of three pods containing magnetic elements from the planets Cryptos and Phantos, plus the moon Nula. Each are connected to receptors in the ship's wings which draw power from their planets of origin in conjunction with the centrally placed Los Crystal.

35 Pod bay inner airlock.

36 Compressed air tanks and pumps for double airlock system.

37 Secondary airlock leading to pod bay.

38 Outer access doors.

39 Airlocked personnel hatch also provides access for the external inspection pod to and from the vessel.

40 Port wing engine nacelle.

41 Tengam drive magnetic element power receptor panels.

42 Tengam/ion drive power processing chamber.

43 Ionisation chamber.

44 Ion particle engine powered by Tengam drive brings vessel up to – but not exceeding – light speed.

45 One of four rocket motors in each nacelle used for take-off, landing and manoeuvring.

46 Rocket engine nozzle heatsink outlet ports.

47 Port side facing retro rocket nozzles, four on each side of the nacelle.

48 Super strength support spars maintain wing integrity when the ship lands vertically. The weight of the ship can be supported safely on the three wing tips.

49 Tengam drive power receptor support spars.

50 Wing tip high strength construction supports ship's weight when landed.

51 Wing-mounted rocket fuel tanks.

18 Suspa-cell suspended animation units used by Earth personnel.

19 Suspa-cell monitor systems.

20 Single-occupancy escape pods.

21 Escape pod deployment airlocks.

22 Disintegrator cannon installed by ISF.

The Zylbat was given to Colonel Dare by the Zylans in thanks for solving the problem of the nearby Vort world, and as a replacement for his Earth craft that was destroyed. Being an alien interstellar craft, it was found to have a number of advanced features when it was examined by Earth experts, including a radar that could detect the Mekon's magnetic gas clouds, an autopilot that could avoid them, and a magnetism-proof hull alloy. Needless to say, these features were subsequently duplicated on Earth ships.

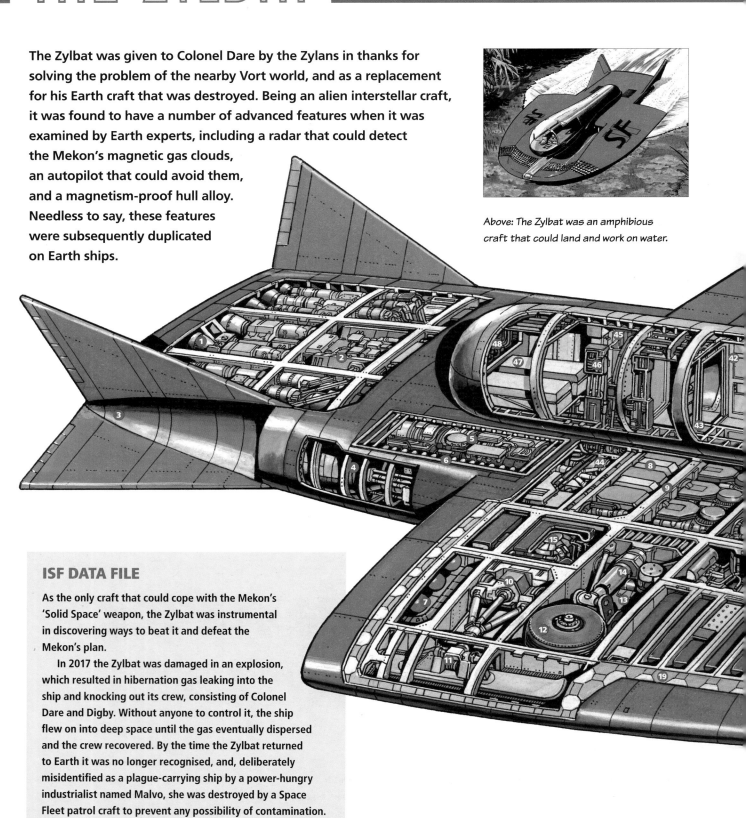

Above: The Zylbat was an amphibious craft that could land and work on water.

ISF DATA FILE

As the only craft that could cope with the Mekon's 'Solid Space' weapon, the Zylbat was instrumental in discovering ways to beat it and defeat the Mekon's plan.

In 2017 the Zylbat was damaged in an explosion, which resulted in hibernation gas leaking into the ship and knocking out its crew, consisting of Colonel Dare and Digby. Without anyone to control it, the ship flew on into deep space until the gas eventually dispersed and the crew recovered. By the time the Zylbat returned to Earth it was no longer recognised, and, deliberately misidentified as a plague-carrying ship by a power-hungry industrialist named Malvo, she was destroyed by a Space Fleet patrol craft to prevent any possibility of contamination.

1. Chemical rocket thrusters.
2. Fuel feed distributor.
3. Heat sink finial.
4. Main drive combustion chamber.
5. Starboard main drive: Zylan-originated magnetic turbine and particle accelerator.
6. Starboard inspection hatch and cooling vent.
7. Starboard wing-mounted air tanks.
8. Ion generators.
9. Retro rocket strengthened bracing wing spar.
10. Starboard landing ski mounting spar.

15. Starboard wing-mounted VTOL thruster.
16. Rear amid ships VTOL thruster.
17. Starboard manoeuvring/braking thrusters.
18. Retro rocket gate seals.
19. High impact-resistant fuel tank protection layer.
20. Emergency battery maintains life-support systems in the event of a power failure.
21. VTOL thruster combustion chamber.
22. Flight deck gravity plating power unit.
23. Forward amid ships VTOL thruster nozzle.
24. Starboard wing-mounted fuel cells.
25. Nose wheel rotated 90 degrees in stored position.

29. Forward landing ski enables Zylbat to land on liquid surfaces in conjunction with VTOL thrusters.
30. Avionics bay.
31. Toughened solar energy panels.
32. Avionics and instrument panel shroud.
33. Dual-control pilot seats (shown before refitted with horizontal pilot's couches).
34. Retracted searchlight hatch.
35. Cabin lighting array.
36. Astral-navigation console screen.
37. Video communications console.
38. Forward scanner computer console.
39. Port atmospheric braking flap.
40. Coma-gas hibernation control valve.
41. Port side crew bunk.
42. Food preparation and storage unit.

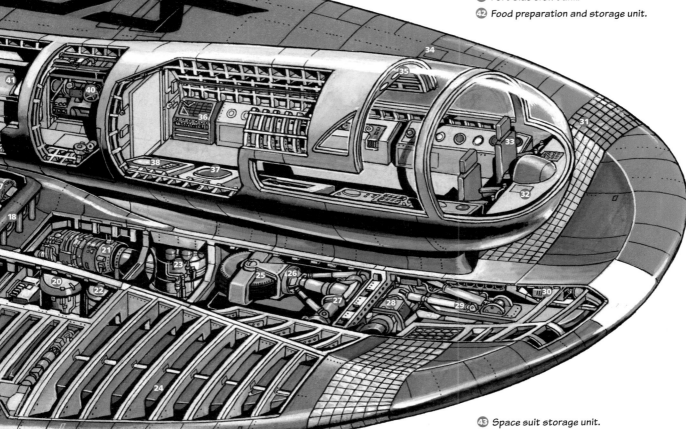

11. Starboard wing-mounted landing hydrofoil.
12. Starboard main wheel.
13. Main wheel door.
14. Telescopic main wheel strut.

26. Nose wheel rotation motor.
27. Nose wheel strut support brace.
28. Dual-system nose wheel and ski/hydrofoil deployment motor.

43. Space suit storage unit.
44. Underside particle beam weapon generator.
45. Shower, basin and toilet cubicle.
46. Hibernation and coma gas control systems.
47. Cryogenic crew hibernation cabinets for interstellar travel.
48. Life-support system.

OTHER CRAFT

COSMOBE CLUSTER SHIP

These massive interstellar craft passed through the Solar System in search of a new home. Their Cosmobe crews are highly intelligent amphibious creatures that need to live in water in order to survive, so their craft contain lots of water-filled spaces. When they encounter other races they secure their spacecraft to the cluster ship's structure with a web of guided grapnels. The crews are then invited aboard to negotiate the possibility of establishing Cosmobe settlements on their worlds, but are not harmed. In planetary atmospheres the cluster ships are capable of disassembling into a large number of smaller powered craft that can fly, but serve mainly as underwater homes.

Left: A Cosmobe cluster ship.

ISF DATA FILE

When the Cosmobe fleet arrived in the Solar System in 2012 all electronic space equipment was jammed by a signal caused by their interstellar drive.

Just one Cosmobe ship remained behind to set up a colony on Earth, while the rest of their fleet moved on in search of other suitable star systems.

PESCOD INTERSTELLAR SHIP

These aggressive interstellar craft came from I-Cos, a dying star system. Being a race of fish-people, they search out planets with oceans, and don't care who they have to displace in order to settle them. Their main weapon – which works equally well in outer space, planetary atmospheres and water – is a red fluid that is sprayed at any threatening object. Known as the 'Crimson Death', this fluid dissolves most solid objects. In order to survive an encounter with it, all equipment needs to be encased in a film of rubber.

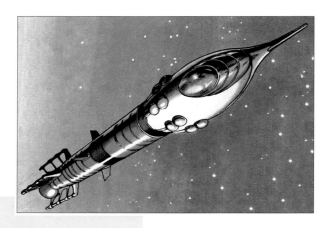

ISF DATA FILE

The deadly Pescod fleet arrived in the Solar System in 2012, destroyed all in-space defences in its path and took to the oceans of Earth. They then decided to establish a base by using the 'Crimson Death' to drill into an underwater cliff face, without realising that it was a dormant volcano. When their drill hit a sensitive area the volcano erupted in a violent explosion, destroying the Pescod fleet in the process.

Above: A Pescod interstellar craft.

Below: The Crimson Death weapon attacks solids like acid – explosive decompression does the rest.

UNIDENTIFIED HOSTILE ALIEN CRAFT

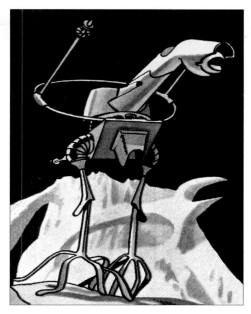

In 2016 an unknown hostile race appeared in the Solar System. They captured Nimbus 1, abandoned its crew in space, and took over a deserted mine on Amalthea, a moon of Jupiter, to be their base. Their strange-looking craft can outpace everything ISF has apart from photon- or Halley-drive spaceships, and are armed with ten spot blisters (five on each side) that shoot white fireballs. These fireballs act like micro-meteorites, eroding the skin of whatever they hit until it fails and the craft depressurises. This enables the aliens to capture ships almost intact.

Unfortunately the mother craft appears immune to Earth weapons – the only way this one got stopped was by turning the full blast of a Nimbus drive on it. Regrettably the craft then exploded, and the secrets of its technology were lost.

Other items of the aliens' equipment were encountered at the mine on Amalthea, including two different sentry robots, a shuttle spacecraft and handguns. The exterior guard robot fired orange bolts at anything that moved, while the interior one first sprayed a white, sticky, immobilising liquid at intruders, then moved in with three cutting arms to dispose of them. Space Fleet handguns were ineffective against the robots. There was no chance to examine the shuttle, as the aliens took it back to their mother ship along with all their other equipment.

Should any of these aliens be encountered, it is advised that ISF be informed immediately.

ISF DATA FILE

The only piece of alien equipment the ISF recovered is the handgun Colonel Dare liberated at the mine on Amalthea.

Top: Exterior sentry robot.

Above: Interior sentry robot.

Left middle: The alien craft firing.

Left: The aliens' mother ship.

THE WANDERING WORLD

The Wandering World is the strangest alien spacecraft encountered so far. Its outward appearance is that of a giant ball of dirty bubbles, propelled through space by three massive radioactive thrust jets placed around it.

The foam bubbles insulate the craft beneath them from outer space, and are also employed to dispose of waste products – unwanted bubbles break away and drift off into space, where, without the continued surface renewal provided by the main mass to sustain them, they deteriorate and eventually burst, expelling their contents into the void. The bubbles also have a strong skin able to withstand impacts or the mass of objects or personnel landing on them. However, any movement causes the bubble to rotate inwards, spilling whatever is on its surface into the craft beneath.

Below the bubble canopy is a multilevel artificial world of linked bubbles, built around a metal core. A variety of plants and creatures live within it, overseen by a slim, large-eyed, blue-skinned race known as the Navs. The craft is powered by a great nuclear reactor, but its gravity is created by centrifugal force as it spins on its axis. The Wandering World is therefore the reverse of Earth – going down takes you nearer to the outside, while heading up takes you towards the centre of the ship.

ISF DATA FILE

The Mekon found refuge in this world in 2018 and tricked the Navs into helping him. However, when Xel and Earthmen arrived the resulting turmoil was too much for the Navs, who ejected all of them into space and continued their journey.

PITTAR SCOUT SHIP

This gigantic craft is small when seen alongside Pittar capital ships, but even so, it dwarfs Space Fleet's new Galaxy-Class battleships. These are the only pictures of them after the only one encountered was struck by a Star Fleet nuclear missile.

It is advised that Space Fleet command be notified immediately if Pittar ships are encountered.

ISF DATA FILE

Luckily the Pittars showed no interest in the inner planets of the Solar System, which were perhaps too small for them, and they have now fled, as it turned out that they were badly affected by Earth's common cold virus. This is just as well, as Earth forces would have difficulty trying to stop a fleet of them or their bigger ships if ISF's nuclear missiles potentially created very little damage.

Left and below: Two views of the damaged Pittar scout ship.

Right: Two Galaxy-Class battleships streak past the giant Pittar scout ship.

TRITON CRAFT

The wrecks of several Triton craft were recovered after their invasion of Earth was thwarted. There was great interest in these – it was known that Xel was the driving force behind their construction, and it was hoped that they might have new or innovative equipment on board. However, they proved to be technologically disappointing – whether because of a lack of resources on Triton, or a lack of time to construct them, it will never be known.

Their spacecraft resembled enormous spinning tops propelled by old-fashioned rockets, with missile pods spaced around the circumference. Their performance was not exceptional, but when employed in large numbers it needed the combined strength of the Earth and Theron defence fleets to stop them.

There were two types of Triton craft: a large, lumbering transport, and smaller, faster, more elongated attack craft.

Right: The Tritons' fast attack craft.

Above: The Triton invasion fleet take off. Four missile pods are mounted around the circumference of Triton invasion craft.

THE RED SHIP

This massive ship was found in the Sargasso Sea of space, where it remains. It is off limits to all but official research teams. Little information has been released about it, except that such craft are extremely dangerous. All communication is blocked in the vicinity of the ship, after which it deploys tiny, deadly attack ships to defend itself. Its crew consists of the aggressive Krevvid race, who are capable of lying dormant for years.

The ship is included here for recognition purposes. Space Fleet command should be notified immediately if such a ship is encountered.

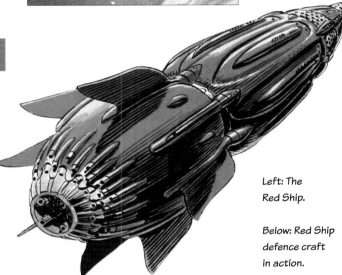

Left: The Red Ship.

Below: Red Ship defence craft in action.

ISF DATA FILE

The Red Ship was found in the Sargasso Sea of Space in 2002, by SF officers King and MacFarlane. They were unable to gain entry, but in 2012 debris hitting the craft after an MH explosion brought it to life: a pod opened and waves of hostile craft spewed forth to defend the ship.

Research teams later boarded the vessel via the open pod bay, and encountered serious difficulties (details remain classified) and a new hostile race – the Krevvid.

NEW DRIVE SYSTEMS

M.H. DRIVE

The tremendous latent power of monatomic hydrogen (or MH) held the promise of a superfast space drive, which was needed to explore the outer Solar System; but it was unstable and likely to explode at the slightest provocation. After a number of accidents Space Fleet's research team thought that rendering it into a jellified form might stabilise it, and in 1997 they built a

Left: The remains of MHX-2 are discovered.

ISF DATA FILE

In 2012 the remains of the missing 1998 experimental MH ship MHX-2 were found in the Sargasso Sea of Space, but it exploded soon after being boarded.

new test ship, which exploded just after take-off. A year later the scientists tried again with a modified version. The test ship's acceleration was fantastic ("gone in the blink of an eye") but that was the last anyone saw of it for many years. After this Space Fleet dropped the project; too many lives were being lost.

Two years later a civilian scientist, Doctor Blasco, found a way to stabilise monatomic hydrogen by beaming special radio waves into it – he called them 'lockwaves'. However, this only worked if the lockwaves were transmitted at exactly the right wavelength, and for some reason once the ship was in motion the required wavelength kept changing. To get around this, pilots must fly with a lockwave tuner, and constantly retune the transmissions in order to keep the fuel stable – which is not as easy as it sounds, as two different lockwaves have to be retuned at the same time. Two-man training ships were built to familiarise pilots with the new technique, three of which were written off in the process, but once mastered the system worked, and it was installed in the specially-built, superfast, super-armoured deep-space vessel Valiant.

ROTOR CRUISERS

After the problematic early attempts to explore the asteroid belt, in 2001 a new type of spacecraft was designed to take up the challenge – the rotor cruiser. As with the ships used in the early days of space travel, since these ships would be out of range of impulse waves for years at a time they were designed to coast most of the way, and were atomic-powered. They had four medium-power main drive engines, but could run on just one if necessary; having four gave it more thrust when needed and provided a useful safety margin in case of any failures.

Being a special exploration craft, scientists didn't want gravity plate technology on board in case it disrupted delicate instrument readings, so the ships used centrifugal-force gravity instead. This is achieved by spinning the outer hull around a zero gravity axis, hence the technical name of such craft – rotor cruisers.

Left: A rotor cruiser found in the Sargasso Sea of Space.

PROTO DRIVE

When Earth engineers examined Lero's Crypt ship in 2001 they did not find the Tengam drive, but they did take note of other design ideas, which in turn led them to develop a supercharged upgrade of the traditional impulse system, called 'Proto drive'. The first type of ship built to this specification was the Z-19 class, which was undergoing trials when the Mekon's robot invasion fleet struck in 2002. A Z-19 test ship survived the war and after being stranded in the Sargasso Sea of Space was recovered in 2012. It proved that the enhanced impulse drive did indeed live up to expectations, and Space Fleet's designers immediately began to install it in other ships.

Above: The Z-19 salvage operation in the Sargasso Sea of Space in 2012.

ISF DATA FILE

XC-9 Space Cruisers were built around a Proto-drive system, and the impressive speed improvement triggered a programme to upgrade all interceptors to this type of power.

HALLEY DRIVE

In 1972 Scottish scientist Halley McHoo accidentally discovered the multiple spin states of tachyon particles. From this discovery he developed an extremely powerful new method of spaceship propulsion – the Halley drive. Tested satisfactorily in 1980, he then went on to build a two-man starship in 1983. This was the Galactic Pioneer, which was launched from the McHoo asteroid base towards Terra Nova, a planet in a distant star system. Unfortunately most of the base was destroyed in a tremendous explosion seconds after the launch, and Halley was killed. His son Galileo McHoo spent the next 30 years rebuilding the base, piecing together his father's work and building an enormous new interstellar craft, the Galactic Galleon, to fulfil his father's dream.

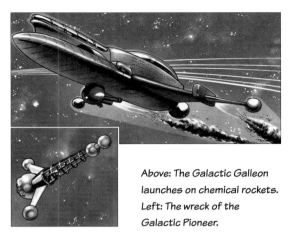

Above: The Galactic Galleon launches on chemical rockets. Left: The wreck of the Galactic Pioneer.

In 2013 the Galactic Galleon became the first man-made craft to voyage to another star system and return safely.

Left: The Galactic Galleon using Halley drive.

ISF DATA FILE

The Galactic Galleon found the remains of the Galactic Pioneer along with the body of its co-pilot Copernicus McHoo in orbit around Terra Nova.

PHOTON DRIVE

The idea of using photon particles to propel a ship through space had been talked about for a long time, but it was only when a really efficient particle gun was developed, capable of firing the particles in an ultra-concentrated beam, that the idea became a reality. The Nimbus craft exceeded all expectations, and was to give mankind a drive system that could quickly reach the outer planets of the Solar System.

ISF DATA FILE

Photon drive made it easy to maintain regular contact with the Saturnian moons. Unfortunately, however, this was looked upon by some Thorks as a threat, and caused tension between the two races.

TIME TRAVEL

Successive expeditions to the Sargasso Sea of Space yielded a number of alien spacecraft, believed to have been deposited there as a by-product of the Red Moon's destruction. Although damaged, one of these contained equipment never seen before, and scientists eventually deduced that it must be a time displacement device. It was some years before a team of scientists and technicians led by Colonel Wilf Banger were finally able to build an experimental ship – the Tempus Frangit ('Time-Breaker') – in 2019.

Left: A Nimbus spacecraft under boost.

Right: The Tempus Frangit.

M.H. TRAINING CRAFT

Built only to teach pilots how to handle lockwave tuners, these two-man training craft basically consist of two MH fuel tanks attached to an engine and a lockwave tuner, with pilot and lockwave operator escape capsules mounted on top. The little ships do not have any landing gear as they are designed to parachute into the sea, where they will float until retrieved. If the fuel becomes unstable in flight a red light comes on, and the crewmen then have only seconds to eject. The escape capsules themselves give the specially space-suited crew extra protection while in space, but once in a planetary atmosphere can be split open at the touch of a button, and the crew then parachutes to safety.

ISF DATA FILE

Three craft were lost in training, but one was used by Colonel Dare to rendezvous with the Valiant in space, and later to escape from the captured spaceship Phobe, one of Saturn's moons.

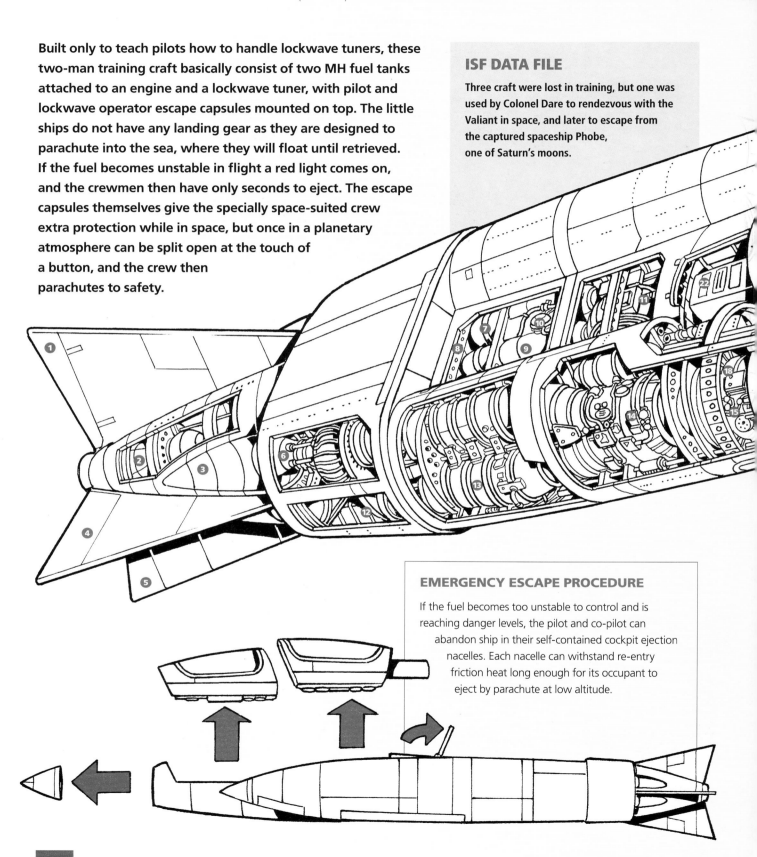

EMERGENCY ESCAPE PROCEDURE

If the fuel becomes too unstable to control and is reaching danger levels, the pilot and co-pilot can abandon ship in their self-contained cockpit ejection nacelles. Each nacelle can withstand re-entry friction heat long enough for its occupant to eject by parachute at low altitude.

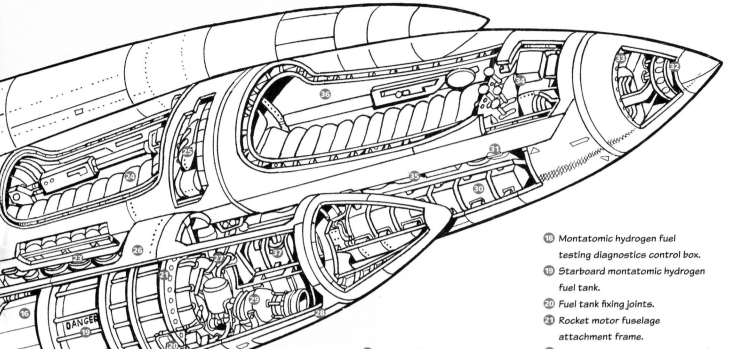

1 Upper tailfin rudder, used in atmosphere.

2 Atmosphere emergency braking parachute.

Below: Escape pods away.

3 Inverted rocket exhaust cowling.

4 Starboard atmospheric aileron.

5 Lower tailfin rudder.

6 Upper starboard rocket exhaust nozzle.

7 Aft air tanks.

8 Rocket nozzle and aft air tank fairing.

9 Air tanks.

10 Life-support and air-recycling systems.

11 Air pumps and conduits leading to gate-sealed cockpit valves; closed if cockpit nacelles need to be ejected.

12 Rear starboard airtight flotation tanks enable the craft to land on water.

13 Secondary twin heat exchangers.

14 Primary rocket motor heat exchanger.

15 Lockwave MH fuel stabilisation ring.

16 Helium pressurisation gas tank maintains montatomic hydrogen fuel stabilisation.

17 Fuel tank maintenance access panel.

18 Montatomic hydrogen fuel testing diagnostics control box.

19 Starboard montatomic hydrogen fuel tank.

20 Fuel tank fixing joints.

21 Rocket motor fuselage attachment frame.

22 Rear cockpit ejection survival pack.

23 Rear cockpit ejection rockets.

24 Co-pilot cockpit.

25 Lockwave tuner control console.

26 Fuel filter cap.

27 Lockwave power and control conduits.

28 Starboard retro rocket operating door.

29 Starboard retro rocket.

30 Forward airtight flotation tanks enable the craft to land on water prior to recovery.

31 One of four cockpit nacelle attachment jacks.

32 Forward emergency braking parachute.

33 Avionics bay.

34 Ergonomically simplified flight control panel.

35 Forward cockpit ejection rockets.

36 Pilot's ejection cockpit nacelle.

37 Standby air pump enables compressed air from flotation tanks to be pumped into pilots' cockpits in the event of an emergency whilst still in space.

THE VALIANT

In 2000 the Valiant was the fastest and most powerful Earth battlecraft ever built. Armed with mobile multi-gun space cannon batteries that retract for take-off, missile ramps and a centrifugal spinning mortar that flings time-fused bombs out into space, she was formidable. Launched from a special enclosed tower like a bullet, once in space the outer shield fell away in two pieces, her hatches folded back and her defence guns deployed. At destination she landed tail first on supporting legs.

Right: The firing shield falls away.

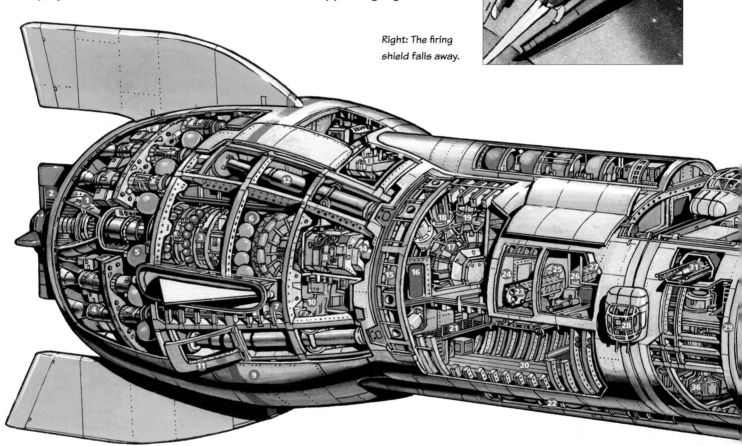

Below: The Valiant in action.

ISF DATA FILE

The Valiant was hastily built to investigate Saturn, as hostile 'Black Cat' craft were emanating from that area. She was captured on the way by her rogue designer, Dr Blasco, but nevertheless completed the journey successfully. During the Titan war Space Fleet regained control, but in her pursuit of Blasco's craft the Valiant's fuel became unstable and had to be dumped, and the drives from Saturnian craft had to be used to get her home. No more MH craft have been built since, as the fuel is now considered far too dangerous to use.

1. Reactor heat dissipation cone.
2. Graphite reactor cooling vanes.
3. Nuclear reactor provides power for heat, light, life support, gravity generation and lockwave fuel stabilisation systems.
4. Chemical booster rockets use both standard rocket fuel and more efficient – but less stable – monatomic hydrogen propellant.
5. Chemical rocket fuel tanks.
6. Helium pressurization gas tank ring serving adjacent monatomic hydrogen propellant tanks.
7. Starboard monatomic hydrogen tank.
8. Lockwave fuel stabilisation generator ensures the helium-pressurised monatomic hydrogen fuel remains within specified safety parameters.

15. Rear access hatch to engine room.
16. Forward access hatch to engine room.
17. Centrifugal mortar automated rotation and bomb-loading assembly.
18. Time-fused mortar bomb.
19. Centrifugal mortar cannon.
20. Cargo hold.
21. Cargo hold loading supervision gallery.
22. Water tanks.
23. Rear starboard space cannon battery.
24. Decompressed space cannon bay.
25. Pressurised cannon operator capsule.
26. Central corridor.
27. Port space cannon battery.
28. Missile ramp turret on hull guidance slot.
29. Grooved turret running track.
30. Starboard missile ramp bay door.
31. Unmanned centrally mounted starboard missile launcher.

36. Artificial gravity generator; powered by reactor at the rear of the craft, providing localised and variable gravity to Valiant's deck plating after lift-off.
37. Compressed air tanks.
38. Port side space cannon battery bay.
39. Crew sleeping bunks.
40. Deck gravity plating.
41. Docking bay.
42. Lower observation deck and lounge.
43. Observation deck
44. Acceleration couches, used by all the crew when the Valiant takes off from an upright position. Only seven couches are shown; up to 25 are used by personnel on take-off.
45. Airlocked docking bay access corridor.
46. Forward starboard retro rocket.
47. Retro rocket fuel tanks.
48. Avionics and forward sensor array.
49. Radar scanner.
50. Pilot's couch with video screen.
51. Lockwave tuner/monitor system.
52. Life support and ship's systems console.

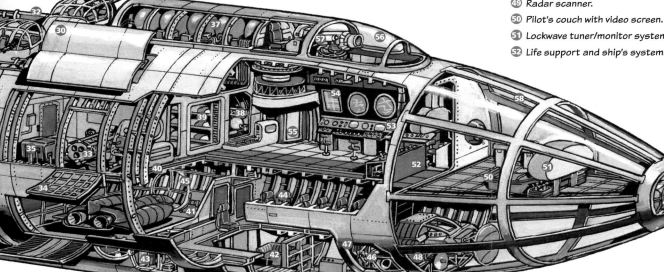

9. Monatomic hydrogen fuel tank emergency ejection hatch.
10. Engine room.
11. Landing leg door.
12. Telescopic landing leg in retracted position.
13. Life-support and recycling systems.
14. Centrifugal motor rotation arm, located within the lockwave generator and powered by the reactor, controls the Valiant's mortar system.

32. Missile ramp bay weapon's system. After two gunnery officers have boarded each airtight turret, the ramp bay is decompressed. The bay doors both slide forward and aft, allowing each turret to move around the Valiant's hull.
33. Forward starboard space cannon.
34. Space cannon battery hull door.
35. Armoured space cannon magazine.

53. Camera output monitor screen.
54. Radar and sensor screens.
55. Access door to observation dome.
56. Observation dome; polarised, heat and radiation resistant.
57. Long-range telescope and camera; can be operated manually within observation dome, or automatically from the control room.
58. Toughened laminate nose-cone bonded with ceramic additives, polarised to reduce solar glare and solar radiation resistant.

THE VALIANT   **111**

ROTOR CRUISER

There were high hopes for the two rotor cruisers that were completed early in 2001 (see page 106 for details), but conviction waned after South American dictator Donanza stole one and wrecked it. The surviving craft did make a proving trip to the edge of the asteroid field, but when a number of systems presented problems confidence failed altogether, and following its return it was decided not to risk lives on a prolonged mission. Instead, another drawing board design for exploring the asteroid field received the green light, and the Marco Polo was built.

Left: Rotor cruiser adrift in the Sargasso Sea of Space.

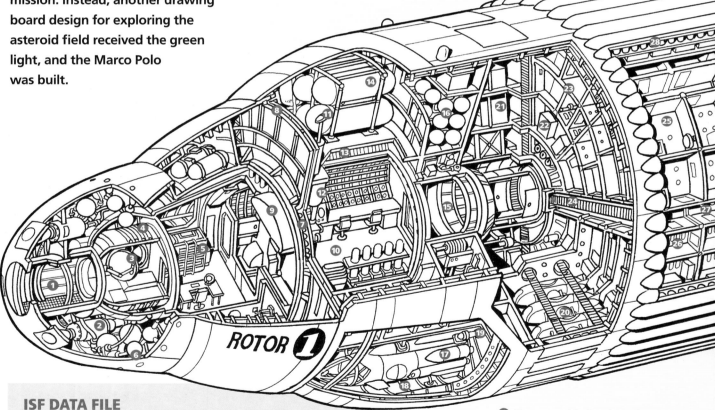

ROTOR 1

ISF DATA FILE

Two rotor cruisers were built, but one was stolen from Mars 2 space station before its mission started. During the ensuing pursuit it was damaged by meteorites, the crew captured and the impaired craft abandoned. It was later found in the Sargasso Sea of Space, and in 2011 Colonel Dare used it to refuel his spaceship Anastasia to escape back to Venus.

1 Nose cone airlock and mooring tube docking bay.
2 Bow retro rockets.
3 Inner air locked entrance bay.
4 Entrance tube bypassing Captain's cabin and control room.
5 Captain's duty cabin.

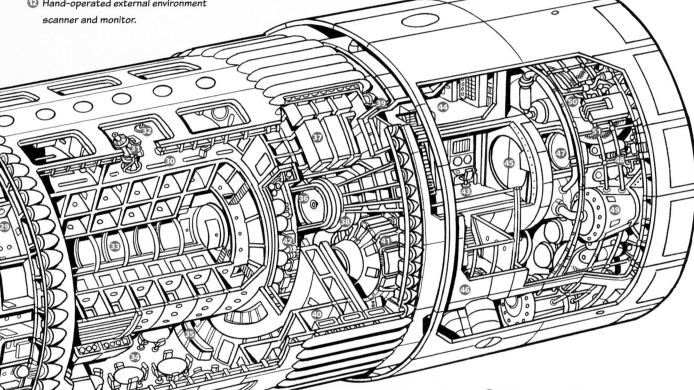

6 Forward pitch and yaw correction rockets.

7 Frictionless gravity field-assisted main body rotation gearing.

8 Forward joint head of fixed and moving hull sections.

9 Primary pilot's control instrument console.

10 Zero gravity control room.

11 External environment sensor and camera.

12 Hand-operated external environment scanner and monitor.

19 Transit clamps prevent damage to landing craft during rotor ship's rotation.

20 Emergency escape pods.

21 Cargo decks and equipment storage.

22 Communications room.

23 Crew accommodation decks.

24 Access ladder to all decks.

25 Officers cabins.

35 Bar and food preparation area.

36 Engineering deck bulkhead door.

37 Cargo decks containing food and water supplies.

38 Aft main body frictionless rotation gearing.

13 Control servicing room with junction and fuse boxes.

14 Air tanks.

15 Zero gravity flight deck central corridor. Crew wear magnetic boots on the steel floor to ensure a sense of gravity is maintained.

16 Air recycling/cleansing and life-support systems.

17 Rotor ship shuttle and planetary landing craft.

18 Landing craft hangar doors.

26 Crew cabins.

27 Cabin access corridors.

28 Experimental cosmic ray collectors.

29 Laboratories specialising in deep-space observation and planetary samples studies.

30 Space observation deck.

31 Deep-space telescope.

32 Telescope operator's seat.

33 Zero gravity central access corridor allows large containers/supplies to be moved around the ship with ease.

34 Crew dining area. Further crew relaxation areas, including a gymnasium, are located on adjacent outer hull decks.

39 Aft joint head of fixed and moving hull sections.

40 Offices, library and crew study rooms.

41 Shielded atomic fission reactor powers all ship's systems as the vessel is designed to operate beyond impulse wave range.

42 Electricity generating ring.

43 Engineering deck.

44 Heat and internal ventilation control station.

45 Bulkhead inner airlock door.

46 Emergency escape/maintenance airlock.

47 Aft outer airlock hatch.

48 Aft pitch and yaw rockets.

49 One of four chemical main drive rockets.

50 Chemical rocket fuel tanks.

51 Oxidant blender tanks.

THE GALACTIC GALLEON

The Galactic Galleon was the first human-made ship to travel safely to a distant star system and back, and do it in style! She was like a small town in space, with luxurious quarters and ship-grown food, even down to a herd of dairy cows for fresh milk. Armed with missile tubes for defence, she also had two scout spacecraft on board. The story of the Halley drive is told on page 107.

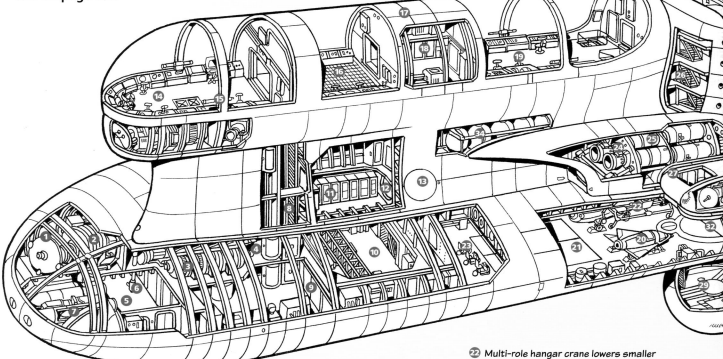

1. Forward space-time distort field projection array.
2. Space-time distort field generator.
3. Time distort field projector coils.
4. Forward generator engineering deck.
5. Missile and sensor probe launch control room.
6. Missile control operative's targeting scanner.
7. Missile launch tubes; also used to launch unmanned sensor probes or even rescue equipment pods.
8. Lift and steps to all decks.
9. Forward laboratory, primarily used for space research and planetary sample analysis.
10. Cargo and supplies storage bay.
11. Spacesuit and hand held equipment storage lockers.
12. Airlock inner door.
13. Port airlock outer hatch.
14. Flight control deck.
15. Communications station.
16. Navigation bay.
17. Sealable bulkhead contains emergency control systems, space suit storage and limited life-support systems and supplies if main control decks are out of action.
18. Emergency supplies lockers.
19. Halley drive control/monitoring station deck.
20. Planetary surface landing ship.
21. Main hangar door enables large craft to enter loading bay.
22. Multi-role hangar crane lowers smaller craft through internal hangar door to underside launch nacelle. It is also used for maintenance and repair of service craft and loading cargo into the hangar.
23. Halley-designed space scooters clamped to hangar bay floor.
24. Retro/manoeuvring rockets.
25. Fuel and oxidant tanks.
26. Access lift and steps to all decks.
27. Port observations nacelle.
28. Long-range telescope, sensor and camera arrays.
29. Underside maintenance vehicle and shuttle launch nacelle.
30. Port side airlocked launch tube.
31. Artificial gravity generator.
32. Internal hangar doors accessing underside launch nacelle below.

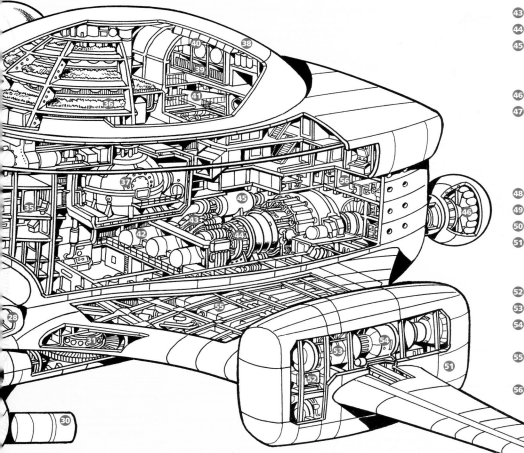

43 Tachyon spin generator.

44 Halley tachyon space-time hyperdrive.

45 Space-time stasis field generator ensures the crew remain unaffected by the ships time-distorting tachyon hyper drive.

46 Space-time vortex stabiliser.

47 Field localiser manifold ensure the distortion field projected around the ship does not affect objects in immediate vicinity as the craft's tachyon drive pushes it through space-time.

48 Space-time distortion generator.

49 Port space-time distort projection array.

50 Wing-mounted generator conduits.

51 Nacelle housing port twin chemical rockets used for initial departures from any given location, before tachyon drive is engaged.

52 Fuel blender pumps.

53 Fuel tanks.

54 Port side chemical rocket combustion chamber.

55 Wing-mounted additional chemical rocket tanks.

56 Commander's quarters.

33 Water recycling tanks and purification systems.

34 Air tanks.

35 Crew quarters.

36 Multi-level hydroponics unit provides fresh food and additional atmosphere cleansing support systems.

37 Nuclear reactor provides power for ship's systems and Halley drive.

38 Life-support dome is the first of its kind on a spacecraft that faces the challenges of interstellar travel. As well as air recycling/cleansing using the largest hydroponics unit on a spaceship ever built, use is made of agricultural systems that include livestock as well as plant production.

39 Livestock holding pens; animals including cows, chickens and pigs are sedated during accelerated space-time journeys.

40 Carbon dioxide and methane reprocessing tanks.

41 Sectioned livestock grazing area.

42 Acceleration inertia compensators.

Below: Front view of the Galactic Galleon

THE GALACTIC GALLEON   **115**

THE NIMBUS

The Nimbus craft were the first Earth ships to use photon drive. This proved to be a highly successful propulsion method, enabling ships to quickly accelerate to speeds of thousands of miles per second. Consequently Space Fleet now had a fast and efficient way to get to the outer planets, and the whole dynamic of the Solar System changed.

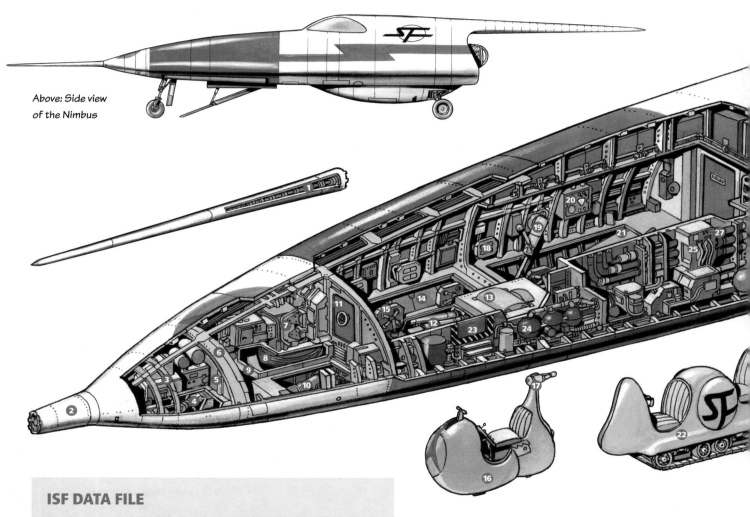

Above: Side view of the Nimbus

ISF DATA FILE

On its very first flight Nimbus 1 was captured by an unknown alien race. Back on Earth no one knew what had happened to it, or if the new photon drive was the cause of its disappearance. Colonel Dare took a volunteer crew out on the Nimbus 2 to find out what might have happened. He discovered that there was nothing wrong with the drive, and subsequently located and saved the crew of the Nimbus 1, then tracked the aliens responsible to one of Jupiter's moons. He clashed with two of them at their experimental base and followed their escape craft back to the mother ship. After being pursued the alien ship was finally destroyed by a full blast from the Nimbus 2's photon jet.

1 Multi-function sensors measure heat, pressure, atmosphere and radiation.
2 Heat-resistant multi-function sensor probe.
3 Sensor probe processing units.
4 Port laser gun (retracted).
5 Avionics bay.
6 Forward pressure bulkhead.
7 Deep-space communications/ data processing unit.
8 Nosewheel, rotated 90 degrees when stowed.
9 Nosewheel stowage well.

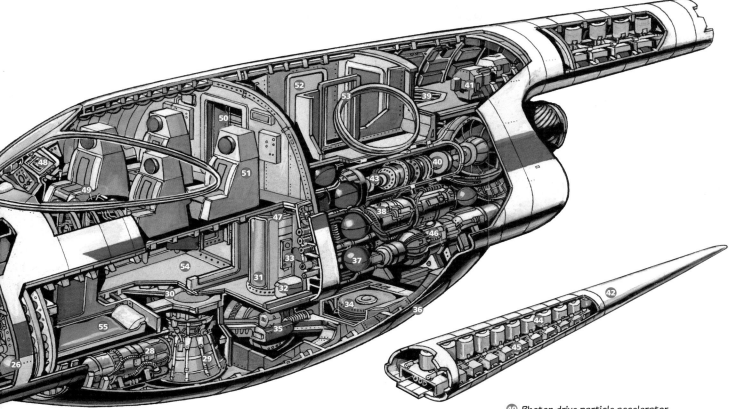

10 Port forward retro/pitch and yaw rocket.

11 Pressurised avionics bay inspection hatch.

12 Nosewheel landing strut.

13 Nosewheel housing.

14 Space scooter door platform in open position.

15 One of four variable-position transit clamps ensure the space scooter is secure.

16 Rescue and maintenance space scooter.

17 Scooter steering rocket.

18 Forward right-hand variable-position transit clamp holds Unitrac in position during take-off and flight manoeuvres.

19 Access ramp deployment hydraulics.

20 Forward bay air pump and pressure control systems.

21 Unitrac access ramp.

22 Unitrac planetary excursion vehicle.

23 Life-support maintenance hatch.

24 Life-support and atmosphere recycling and air 'scrubbing' systems using zyolithic crystals.

25 Fuel injection and regulation management systems.

26 Compressed air tanks connected to forward bay air pump and adjacent life support/recycling systems.

27 Fuel feed lines.

28 Retro rocket.

29 VTOL rocket.

30 High-strength VTOL bracing stanchion.

31 Shower.

32 Toilet.

33 Emergency toilet/shower controls for use in zero gravity.

34 Port landing wheel (retracted).

35 Dual flywheel gyroscope motor assembly.

36 Port landing wheel door.

37 Electricity generating ring linked to adjacent nuclear fusion reactor provides power to life-support gravity systems plus the particle accelerator.

38 Nuclear fusion reactor.

39 Photon drive shielded inspection hatch.

40 Photon drive particle accelerator.

41 Emergency power generator linked to heat generation stanchion's storage cells.

42 Heat generation stanchion: excess heat from photon drive exhaust is gathered by underside sensors linked to storage cells within the fuselage. Stored energy provides additional power and heat.

43 Photon drive particle gun.

44 Photon drive excess energy storage cells.

45 Excess heat collection sensors.

46 Port chemical rocket.

47 Water tank.

48 Photon drive control console.

49 Engine throttle levers.

50 Access ladder from control deck to lower deck via the central corridor.

51 One of four acceleration couches.

52 Starboard airlock inner hatch door.

53 Port airlock outer hatch door.

54 Central access corridor.

55 One of four crew bunks; two each side of central corridor.

Colonel Wilf Banger's extraordinary Tempus Frangit is still in its experimental stage, and so far it tends not to arrive at the exact coordinates intended. More research and development is ongoing.

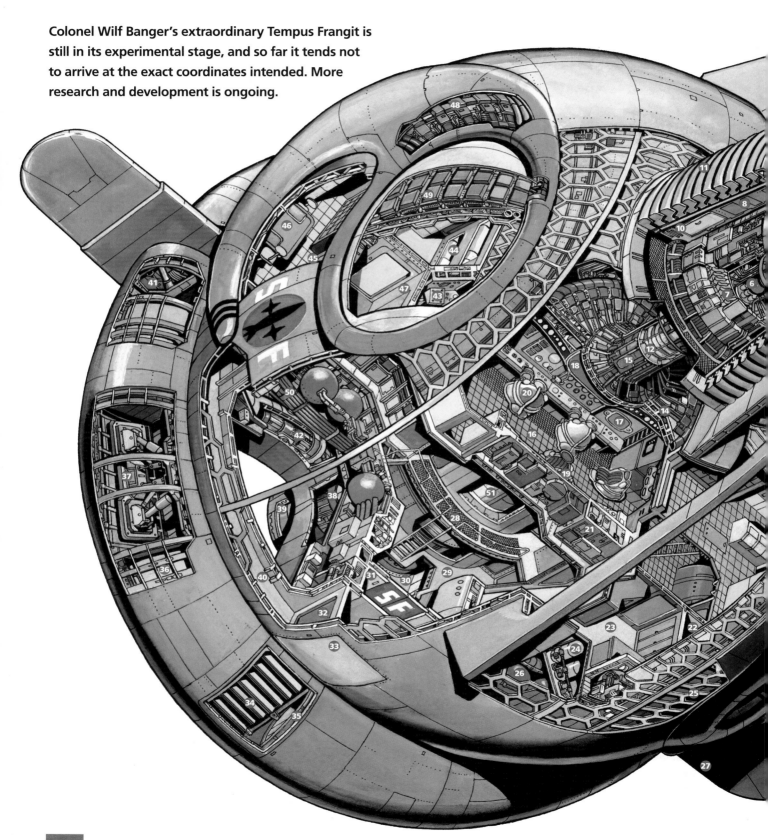

1. Subspace communications antenna.
2. Forward long-range sensor array.
3. Space/time navigation array and homing beacon.
4. Solar energy power converters.
5. Space/time distortion field generator: the Tempus Frangit's interstellar hyper-drive used in conjunction with Photon drive reactor below.
6. Hyperdrive Tachyon core.
7. Time distort field projector coils.
8. Exit shaft to topside hatch.
9. Top exit hatch.
10. Access door from exit shaft to generator maintenance gallery.
11. Time/space distortion field generator projection vanes.

12. Space/time stasis field generator ensures that the crew remain unaffected by the Tempus Frangit's time-distorting hyper-drive.
13. Solar energy collector panels.
14. Artificial gravity generating ring.
15. Magnetic turbine.
16. Control room.
17. Short-range astro screen.
18. Control room video screen.
19. Long range astro-scope.
20. Crew acceleration couches; during take-off these remain in fixed positions, but in flight can be moved around the magnetised floor by each operator if the ship's acceleration prevents crew from walking around.
21. Primary computer core.
22. Space suit storage locker.
23. Stores and supplies bay.
24. Atmosphere recycling and life-support systems.
25. Honeycombe high-strength hull frame construction.
26. Chemical rocket propellant tanks.
27. Atmospheric stabilising fin in deployed position.
28. Reactor maintenance gallery.

29. Electronics and fuse room.
30. Reactor maintenance bay.
31. Inner air locks.
32. Corridor airlock door.
33. Outer airlock door provides access to main corridor in both directions and to inner airlock.
34. Retracted access steps to airlock.
35. Access steps hull door.
36. Air tanks (x2).
37. One of four retracted landing legs.
38. Primary chemical rocket used for take-off and landing.
39. Take-off rocket nozzle.
40. Access corridor encircling the ship.
41. Atmospheric stabilising fin positioning actuator.
42. One of four gimballed manoeuvring chemical rocket thrusters.
43. Geiger counter.
44. Sleeping quarters with access to adjacent hygiene station.
45. Auto chef.
46. Mess room with food preparation and storage facilities.
47. Access well to control room and other levels.
48. Photon particle energy conversion conduits.
49. Port Photon particle collection scoop.
50. Photon particle stream channels.
51. Photon drive reactor.

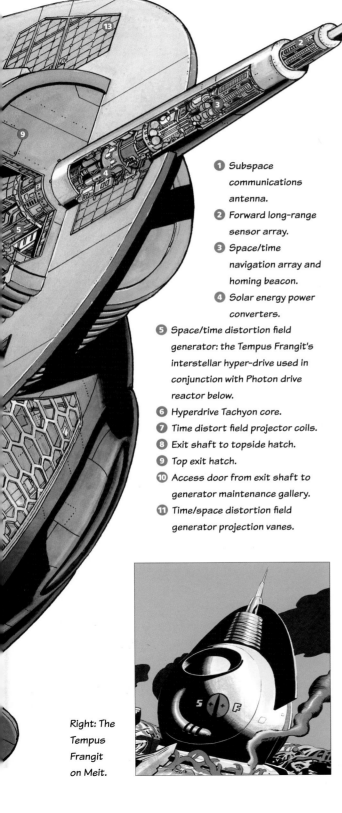

Right: The Tempus Frangit on Meit.

ISF DATA FILE

On its first mission the Tempus Frangit found a twin star system with a single unstable planet, Meit, and encountered difficulties from a hostile stranded alien, a Stoll named Xel. On returning to the edge of the Solar System it encountered the strange Wandering World, with an unwelcome guest, the Mekon, on board. The mission ended with Xel and the Mekon being brought back to Earth as prisoners.

Following modification Tempus Frangit jumped to the Vega system to investigate a strong power source that had been detected. It returned with a new type of nuclear power after helping free the local people from Vendal oppression.

SPACE FLEET HISTORY

What follows is a list of the files that need to be consulted in order to fully understand Space Fleet history, and Earth's relationship with Venus and its occupants. Each entry provides references and a brief description of the file's contents. Classified and unreliable files have been excluded.

SPACE FLEET HISTORY KEY

ABCFR	*ABC Film Review*	HB	Hawk Books	P	*The Sunday People*
DDA	*Dan Dare Annual*	HBD	Hawk Books,		newspaper, 20 April
DDSB	*Dan Dare's*		*Dan Dare Dossier*		to 4 October 1964
	Space Book	NE	New *Eagle*,	SA	*Spaceship*
EA	*Eagle Annual*		26 August to		*Away* magazine
EV	*Eagle* volume/issues		30 September 1989	TB	Titan Books

Mars 1988
EA2, HB2, TB2
Spaceships keep losing control and communications on approaching Marsport.

Moon Run
EA10, HB2
Crews suffer blackouts on the civilian Moon run.

The Venus Story
EV1/1–EV2/25, HB1, TB 1&2
The record of the first successful Venus landings, and the events up to and including the battles of the Earth–Venus war of 1996.

The Gates of Eden
SA9–SA21
As the Venus food runs commence, and Colonel Dare is presented with the Anastasia spacecraft, strange events occur elsewhere.

Raiders in Space
DDSB
Simon Crowley, a South American space operator, refuses to help the Venus–Earth food run and becomes a raider instead.

Asteroid Rescue
EA1, HB5
The Astral Queen goes to the rescue of two space explorers trapped in the asteroid field.

Operation Triceratops
EA4, HB4, TB7
The operation to move a Venusian triceratops from Venus to the Isle of White interplanetary zoo.

The Red Moon Mystery
EV2/26–EV3/11, HB2, TB3
The arrival of the lethal Red Moon at Mars and subsequently Earth, the struggle to understand its purpose and the battle to defeat it.

Marooned on Mercury
EV3/12–EV3/46, HB2, TB4
The experiences of the Red Moon survivors on Mercury, fighting the Mekon.

Operation Plum Pudding
EA5, HB9, TB11
Escaping convicts take over a spacecraft. The situation is saved by Astral Cadet Flamer Spry.

The Double-Headed Eagle
EA3, HB3
An account of how the first interplanetary Olympics almost came to grief.

Dan and Donanza
DDSB, HBD
A South American dictator steels a rotor cruiser to further his plans of world domination.

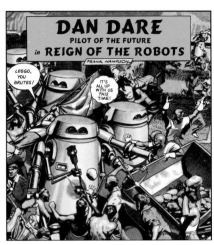

Spaceship Away
ABCFR September 1952, ETV4/4, TB4
Criminals commandeer a spaceship
to facilitate their escape.

Digby the Guinea Pig
DDSB, HBD
Spaceman Digby is volunteered to test
a corium gas inflated spacesuit.

Operation Saturn
EV3/47–EV5/21, HB3, TB5&6
When hostile 'Black Cats' arrive from
Saturn, a mission is despatched in
an MH spaceship to investigate.

Prisoners of Space
EV5/22–EV6/18, HB4, TB7
The expedition to discover what
had happened on XQY satellite space
station, which had broken
off communication with Earth.

Operation Silence
EA6, HB6, TB8
Three planetary leaders are kidnapped
on Venus, but are rescued before this
becomes known.

The Man From Nowhere
EV6/19–EV6/47, HB5, TB8
A spaceship appears from nowhere,
with an alien seeking help.

Rogue Planet
EV6/48–EV8/7, HB6, TB9
A Space Fleet team stops the
aggressive Rogue Planet from
wiping out Crypt civilisation.

Reign of the Robots
EV8/8–EV9/4, HB7, TB10
The fight against the Mekon's
deadly robots, which rule Earth
and Venus.

The Ship That Lived
EV9/5–EV9/16, HB7, TB
An account of how the
Anastasia was saved from
the lava plains of Venus.

The Phoenix Mission
SA1–SA4
The first salvage mission to
the Sargasso Sea of Space.

Green Nemesis
SA4–SA22
The second salvage mission to
the Sargasso Sea of Space.

The Phantom Fleet
EV9/17–EV9/52, HB8, TB11
The Cosmobes and the Pescods
arrive near Earth.

Space Race
EA7, HB10, TB13
Someone attempts to kill
everybody taking part in the
annual race around the Moon.

The Vanishing Scientists
EA9
An Astral College cadet discovers
where vanishing scientists are
disappearing to.

Safari in Space
EV10/1–EV10/18, HB9, TB12
Space pioneers are kidnapped from a holiday on Venus and end up on an asteroid.

Terra Nova
EV10/19–EV10/40, HB9, TB12
The journey to a new world and what is found when the expedition arrives.

Trip to Trouble
EV10/41–EV11/11, HB9, TB13
The battle for freedom on Terra Nova.

Project Nimbus
EV11/12–EV11/28, HB
The Nimbus 1 spacecraft disappears.

The Robocrabs
EA1963, HB5
Alien robot war machines land on Earth

Mission of the Earthmen
EV11/29–EV11/52, HB10
Colonel Dare and Digby help to bring about peace between the warring tribes of Vort, and are given the Zylbat in order to return to Earth.

The Solid Space Mystery
EV11/53–EV12/23, HB11
Pockets of 'Solid Space' disrupt space flight.

The Platinum Planet
EV12/24–EV12/47, HB11
Colonel Dare finds that an entire planet shielded in platinum conceals a helmet-brainwashed civilisation beneath it.

The Earth Stealers
EV12/48–EV13/9, HB11
Hit by climate change and plague, the population of Earth has been evacuated to Mars. Ambitious would-be dictator Malvol plans to seize power on both planets but is defeated.

Mission to the Stars
P, HBD
The Copernicus 1 disappears on its way to Alpha Centauri, and the Copernicus 2 is sent to investigate.

The Solid Gold Asteroid
EA10
Criminal Martin Brand hides gold bullion on an asteroid.

Operation Crusoe
EA1964
The search for a suitable planet by a humanoid refugee ship that is running out of supplies.

Mekon's Revenge
NE 26 August to 30 September 1989
The Mekon captures a ship carrying radioactive waste and seizes Moonbase, from which he plans to contaminate the Earth.

Space Rocks
EA1966
The Mekon causes space rocks to bombard the Earth.

The Planet of Shadows
DDA1963, HBD
Spaceships trying to get to a new planet are surrounded by a strange lethal halo.

The Planulid
DDA1963, HBD, TB7
A alien organism reaches Earth on a meteorite, and threatens to take over the planet.

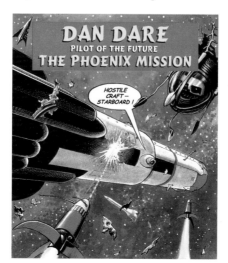

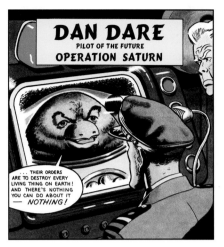

Operation Moss
EA8, HB12, TB13
A deadly dormant life form from
Q3, the planet of Moss, is brought
to Earth.

Operation Earthsavers
EV13/10–EV13/23
A new satellite causes an alarming
increase in plant growth on Earth.

The Evil One
EV13/24–EV13/32
Pursued by an alien fleet, evil
scientist Mal Valus seeks refuge on
Earth, where he has a secret base.

Operation Fireball
EV13/33–EV13/42
A Martian solvent causes a huge,
all-devouring fireball on Earth.

The Web of Fear
EV13/43–EV13/52
Spiders from a mysterious asteroid
leave a trail of destructive webs
across Earth and the Moon.

Operation Dark Star
EV14/1–EV14/9
A dark star and its companion
planet are discovered and investigated
by Space Fleet.

Operation Time Trap
EV14/10–EV14/38
The new Tempus Frangit time
displacement spaceship gets
trapped on the planet Meit,
and encounters the hostile alien Xel.

The Wandering World
EV14/39–EV15/13
The mysterious 'Wandering
World' is found on the edge
of the Solar System. Its population
is found to be controlled by the Mekon
and Xel, who are both captured.

The Big City Caper
EV15/14–EV15/22
Escaping from Space Fleet custody,
Xel starts a teenage rebellion
in London.

All Treens Must Die
EV15/23–EV15/42, HB12
The Mekon escapes to Venus,
planning to replace the Treen
race with his new super-Treens.

The Mushroom
EV15/43–EV16/6, HB12
The Mekon seeds Earth with
hostile artificial mushroom
structures in an attempt to
conquer the planet.

The Moonsleepers
EV16/7–EV16/29, HB12
Xel escapes to Triton and coerces
the locals into building a fleet
to invade Earth.

The Singing Scourge
EV16/30–EV17/6
In the Vega system, the Vendals
enslave the neighbouring Trons
by means of a deadly ray machine
that 'sings' as it kills.

Give Me the Moon
EV17/7–EV17/26
The FIST terrorist organisation
demands the Moon.

The Menace from Jupiter
EV17/27–EV18/, HB12
The Verans from Jupiter seek
Earth's help to free them from
the giant Pittar invaders.

SPACE FLEET HISTORY

 ## SPACE FLEET BADGES OF RANK AND INSIGNIA

Officers' Epaulettes

| Marshal of Space | Area Marshal | Colonel | Pilot Major | Pilot Captain |

Pilot | Pilot Cadet | Area Marshal Engineering | Engineer Colonel | Engineer Major

Engineer Captain | Commissioned Captain | Engineer Cadet | 'Q' Branch: Non-flying and admin

Cap Peaks

Area Marshal and Marshals of Space

Officers

Other Ranks

Arm Insignia

Radio and Radar Technician | Engineer Impulse wave Technician

Engineer Rocket Technician | Engineer Construction

Engineer Electrical Technician | Supervising Engineer all trades

Medals

Comet Medal | Venus 1996 Medal

Cap Badges

Officers | Other Ranks

Pilot's Tunic Badge

FREQUENTLY ASKED QUESTIONS

Is the Mekon really dead?

It has been thought so a number of times, but on every occasion he has subsequently reappeared. Consequently, until we receive physical proof of his demise this cannot be answered with certainty. In the interests of safety we therefore continue to improve our defensive capabilities.

Space Fleet has sent a number of survey ships to the stars. When will colonists follow them?

Though it is true that we have mounted numerous exploratory missions, we would need to be certain of a planet's safety before colony ships could be sent. For example, take planet Q3, nicknamed 'Moss' because that's all that grows there. From this seemingly innocent place puffball spoors were somehow transferred to Earth, and would have wiped out the human race if Spaceman Digby had not brought back a sample of the moss that countered these puffballs. It is for just such reasons that we dare not risk sending colony ships anywhere until such planets have been closely and carefully studied.

Even with Halley-drive ships, journeys to far-flung star systems could take years to get there and back. Is it worth it?

Mankind has always wanted to explore – to discover what's out there is in our nature, and we would not have come even as far as we have if we stopped exploring and investigating each new thing that we encountered. Who knows what may await discovery upon some distant planet that will improve mankind's future?

Won't space explorers making long journeys be old men when they come back?

The crew serve in turn and rotate through suspended animation units, so they don't actually age very much. However, their family and friends on Earth still age physically in the normal way, which is why only volunteers – preferably without family ties – are selected for such trips.

We hear that the Thorks can telesend through space – why cant we?

It is true that the Thorks do have a telesend transport system that can operate in deep space. However, like the Treen planetary system, it is not 100% reliable when it comes to telesending Earth people – one crewman was almost killed using a Thork machine. Consequently until telesend machines are perfected we will not be using them.

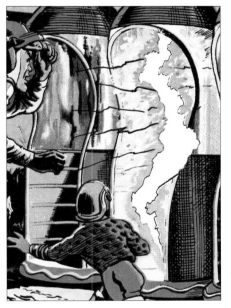

Above: Treen teleports turn some Earthmen upside down at the other end...

...Whereas some Earthmen using Thork teleports almost don't arrive at all.

Is your time-travelling ship the answer to some of these problems?

Our Tempus Fragit time and space travel ship remains susceptible to a variety of teething problems. It does not always go to the exact time or place we would like it to, which means there is a risk that it might become lost and be unable to return. Colonel Banger and his crew are learning more about the ship all the time, and experimental missions will continue to be mounted. Once we understand enough about such ships we may eventually build a fleet of them.

Is there any explanation regarding why we are encountering so many humanoid species on other worlds?

The short answer is no, but the question raises many questions of its own. Did a humanoid race once seed many different star systems, including our own? Or did some long-lost Earth civilisation send out colonies into far space before itself perishing many aeons ago? The universe never ceases to remind us that it is an enormous place full of questions waiting to be answered.

EAGLE

EVERY WEDNESDAY

4½d

COMPANION TO GIRL, SWIFT AND ROBIN

EPISODE 31

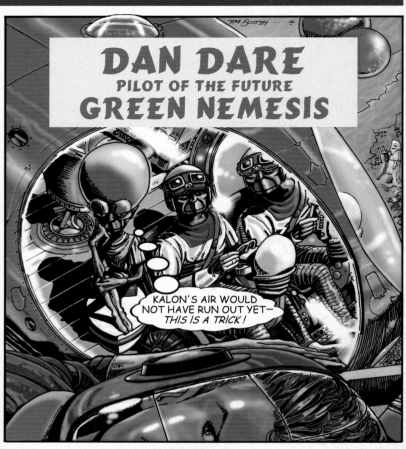

DAN DARE
PILOT OF THE FUTURE
GREEN NEMESIS

KALON'S AIR WOULD NOT HAVE RUN OUT YET— THIS IS A TRICK!